LEEDS COLLEGE OF BUILDING
T02556

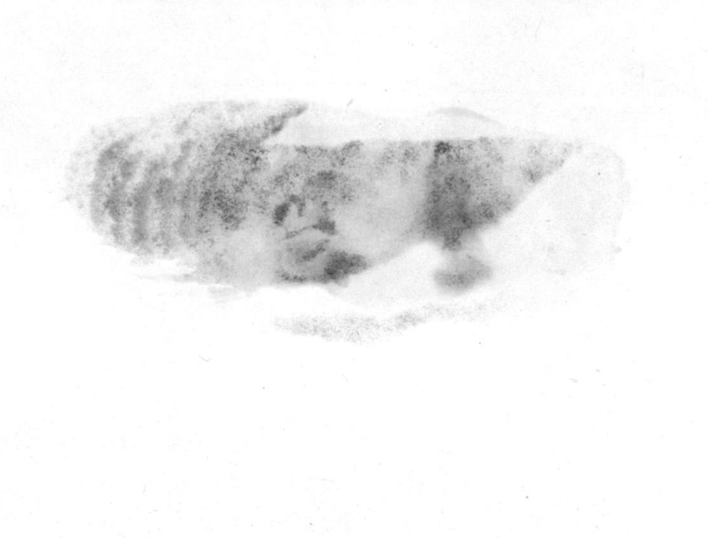

Timber Specifiers' Guide

Understanding and Specifying Softwoods in Building

Timber Specifiers' Guide

Understanding and Specifying Softwoods in Building

Jack A. Baird
CEng, FIStructE, FIWSc

BSP PROFESSIONAL BOOKS
OXFORD LONDON EDINBURGH
BOSTON MELBOURNE

First published 1990

British Library
Cataloguing in Publication Data
Baird, J. A. (Jack Alexander), *1932–*
Timber specifiers' guide.
1. Structural components: Timber. Design
I. Title
624.1'84

ISBN 0-632-02679-0

BSP Professional Books
A division of Blackwell Scientific Publications Ltd
Editorial Offices:
Osney Mead, Oxford OX2 0EL
(Orders: Tel. 0865 240201)
25 John Street, London WC1N 2BL
23 Ainslie Place, Edinburgh EH3 6AJ
3 Cambridge Center, Suite 208, Cambridge MA 02142, USA
107 Barry Street, Carlton, Victoria 3053, Australia

Typeset by DP Photosetting, Aylesbury, Bucks

Printed in Great Britain by
Billing & Sons Ltd, Worcester

The publisher and author have taken care to give information and references from reliable sources which is accurate at the time of publishing but without responsibility on the part of the publisher or author.

CONTENTS

FOREWORD

During my forty odd years of involvement in the development and manufacture of joinery and timber construction products, the poor quality of written specifications and the consequent lack of understanding between specifiers, manufacturers and builders as to what constitutes a specification satisfactory for all concerned, has been a constant source of concern, misunderstanding and dispute. The situation was not helped in the past by impractical British Standards which designers and manufacturers had little option but to disregard in many respects, by the acceptance of poor designs to save on first-cost, and even in the 1960s by the absence of quality paints suitable for exterior use on timber.

Fortunately, in recent years, dedicated individuals have devoted an enormous number of hours, free of charge, to write practical realistic British Standards and Codes on joinery, timber for joinery and wood trim, stress grading, structural and semi-structural items, performance in fire, preservation, and primers and paints for timber, all of which can now have the support of industry. In addition, largely as a result of collaboration between the British Woodworking Federation and the staff of The Swedish Finnish Timber Council, assisted when necessary by BRE, there have been dramatic improvements in the design of components, the availability of exterior quality paints, and the availability of timber graded specially for structural and joinery end uses.

No one has been more involved with this work than Jack Baird. However, there still remains the task of communicating these improvements in standards etc. to specifiers and all those not intimately involved with the improvements. That is why it is so important that this book has finally been written.

The section on typical specifications should go a long way towards improving knowledge on the type of written specifications which are required to allow fair competition, and to give a basis from which specifiers, manufacturers and builders can agree what they really want from what is available. I am delighted that the ten chapters on 'understanding' the properties of timber and associated subjects such as paints have been included and cross-referenced to the chapters on typical specifications. They make this book an invaluable source of reference for anyone involved in designing, specifying, buying, selling, manufacturing, teaching or learning softwoods. Finally, particularly at a time when doubts are being expressed about the future availability of some hardwoods, I am pleased to see the chapters on supply of the main softwoods used in the UK, and to read the encouraging message which they contain.

This book is bound to be an invaluable source of information on understanding softwoods and how to specify them.

Eric Thackray
Technical Director, John Carr Group,
President of The British Woodworking Federation, 1989

INTRODUCTION

A primary object of this book is to assist those who are involved with writing or analysing end use specifications for softwoods used in UK building construction, joinery or wood trim to understand what should be specified, how to specify, and to understand the background to specifications and timber properties.

At the same time, I hope that readers will gain far more from the contents of this book than simply a better grasp of timber specifications. Particularly in the second section which contains background chapters on timber properties, I have aimed to include a summary of most of the knowledge which I have accumulated about timber whilst working with a manufacturer of structural and joinery components, and with The Swedish Finnish Timber Council. I believe, therefore, that the book should be of use and interest to a much wider audience than those involved simply with specifications. I see the contents as including items of use and interest to builders, developers, manufacturers, lecturers and students, and certainly to members of the timber trade including sawmillers, as well as to designers and specifiers.

Having included in the book the type of information which I feel will be useful to a wide readership, I found great difficulty in choosing a title which best describes the contents. In the end I agreed a fairly short title followed by a sub-title. I chose the word 'Understanding' as well as 'Specifying' to convey the general aim in two words. 'Softwoods' and 'in Building' are intended to show the extent and limit of the contents.

The book concentrates on covering use of the following species:

- European Redwood from Sweden, Finland, Russia, Norway and Poland;
- European Whitewood from Sweden, Finland, Russia, Norway, Poland and Czechoslovakia;
- Canadian Spruce-Pine-Fir, Hemlock or Hem-Fir, Douglas Fir-Larch and Western Red Cedar;
- United Kingdom Sitka Spruce etc.;
- Parana Pine from Brazil;
- Maritime Pine from Portugal.

Even though I have omitted to include at this stage the import from some countries such as the USA, New Zealand, West Germany, Belgium etc., which may become more important suppliers to the UK in future, I do not think that this will detract from the main objects of the book.

The book is written to indicate how timber and timber specifications can satisfy British Standards and Codes. So many important British Standards and Codes have been updated in the period of a few years leading up to 1989 that this seems a good time to release the book for printing. I have, however, added the date of standards and codes almost ad nauseam so that specifiers may compare my guidance notes with future amendments and issues by BSI. Of course I am aware of the moves towards the use of CEN Standards but believe that, for building in the UK, reference to the much more detailed and extensive BSI documents will continue to be the most useful basis for specifying for several years to come.

Until writing the book, I had no idea that I would have to refer to over sixty British Standards or Codes. They are an important part of the book and have been brought together in a list at the end of the text for ease of reference.

The book is written in three parts, A, B and C.

The chapters (A1, A2 etc.) in Part A give guidance on how to specify softwoods for individual uses or items. Suggested typical specifications are given in most of these chapters but, even in these cases, I must advise readers to study the background notes in the relevant chapters in Parts A and B and the important relevant BSI documents, make sure that they are up to date, decide for themselves how to present each specification, and decide on the amount of detail which they wish to include.

The chapters in Part B give background reading on the properties of timber which affect its performance. These need to be understood (at least in principle) when specifying. Subjects such as sizes and tolerances, drying and moisture content, stress grading, joinery grading, durability and preservation are obviously important subjects to appreciate. I have also taken the opportunity to include chapters on performance in fire, mechanical fasteners and glued joints, and even notes on painting. The use of an unsuitable paint on the exterior surface of timber can ruin in a few months the performance of an otherwise carefully specified, designed and constructed exterior timber item.

Having given many talks on timber, my experience has been that if part of a talk is not devoted to forestry and the timber supply situation, most of question time is spent assuring the audience of the healthy supply situation for softwoods, particularly from the Nordic area which is the source of about half of the UK import of softwoods. I have included, therefore, in Part C, chapters about the forests and sawmilling production of the Nordic area, Russia, Canada, the UK,

Poland, Czechoslovakia, Brazil and Portugal. In total, these countries provide well over 90% of the UK consumption of softwoods. The reader will appreciate that it has proved difficult to establish the exact situation in some of these countries, but I have tried to do so through recommended trade contacts.

In Part C I have taken the opportunity to include notes on the points which experience shows to be those which regularly confuse non-specialists, such as the correlation (or lack of correlation) between sawmill grades and end use grades, misuse of specialist terms such as 'Upper Gulf', 'Group 1' etc., and how tolerances, re-sawing and other processing affects the size to consider in designs.

Although I have tried to include answers to most of the questions which crop up when softwoods are used for building, I have included very little on timber technology. That subject is covered in many other specialist publications.

Finally, before this introduction becomes a book in itself, may I pay tribute to the many contacts in the timber world, preservation industry, BRE, BSI etc. who have assisted me with various drafts over several years, and also thank my wife Margaret for her patience during the several years it took me to write and re-write the many drafts, and to commend her drawings.

In particular, I would like to thank Peter Grimsdale for his repeated assistance with many chapters and the time he took to study my final draft, and also to thank Ray Hudson. Many members of the timber trade have reason to thank Ray for his desire and determination that they should improve their knowledge of timber properties, and I number myself amongst them.

Jack A. Baird
Lea, Lincs., 1989

TIMBER FOR WINDOWS

A1

INTRODUCTION

A specification for timber in windows should include details of the following items which may vary in the form of the wording used depending whether it is written by a specifier or a manufacturer:

- *The species which is to be used* or a list of the species which are acceptable to the specifier or, at the very least, a note of whether a softwood or hardwood is to be used. Any species used should be 'Suitable' or 'Suitable if Preserved' as tabled in BS1186: Part 1: 1986 on the quality of timber in joinery (see the notes below on Species and Classes).
- *The Class or Classes of timber required in the final assembly* as detailed in BS1186: Part 1: 1986 (see Chapter B4).

  As explained in Chapter B4, it is preferable not to refer to sawmill grading in a specification for timber in windows. Unless carefully worded, such a reference may even contradict the reference to the Class of BS1186: Part 1: 1986.

  Workmanship, including quality of the surface, should be as detailed in BS1186: Part 2: 1987.

  If a non-opaque finish is to be applied, some of the clauses in BS1186: Part 1: 1986, i.e. those referring to:

  — splits, shakes and checks
  — unsound knots, resin pockets
  — insect attack, discoloured sapwood
  — 'making good' by plugs, inserts or filler, or
  — the use of finger joints or laminating

  may need to be amended or amplified (e.g. if a filler is to be permitted it may have to be a particular type). Obviously such amendments to clauses are more important for 'Exposed' and 'Semi-Concealed' surfaces (see 'Surface Categories' in Chapter B4) than for 'Concealed' surfaces. Note that BS1186: Part 1: 1986 makes the point that amendments to clauses will affect cost and may extend the delivery time.

- *The moisture content range at the time of 'handover'.* Although this point is actually covered for exterior joinery if the specification calls for compliance with BS1186: Part 1: 1986, moisture content is so important that it does no harm to emphasize it by repeating the values given in BS1186: Part 1 (see Table B2.2).
- *If preservation is to be used:*
  - *the type of preservative to be used* (usually an organic solvent in the UK for windows: see Table A1.1),
  - *the method of application* (i.e. double vacuum or immersion for organic solvents), and
  - *the level of preservation required* (i.e. the 'performance category') normally by reference to the 'desired service life' (see Table A1.1) as defined in Table 5 of BS5589: 1989. (The 30 year desired service life performance category should normally be considered to be the notional period suitable and adequate for specifying the preservation of windows.)

Table A1.1 Organic solvent preservation processes which satisfy BS5589: 1989 for timber in windows

Species and amenability of the heartwood to preservation[1]	Desired service life performance category and required organic solvent preservation process			
	30 years		60 years	
	Double vacuum	Immersion	Double vacuum	Immersion
European Redwood				
not thicker than 20 mm 'MR'	+	3 min	+	60 min
thicker than 20 mm 'MR'	+	3 min	+	–
European Whitewood 'R'	+	–	–	–
Canadian Hemlock 'R'	+	–	–	–
Canadian Douglas Fir-Larch 'R'	Not normally required to be preserved			

[1] i.e. Resistant 'R' or Moderately Resistant 'MR' (see Table B5.3).
\+ indicates that the species can be preserved to satisfy BS5589: 1989 to the performance category by the organic solvent process indicated. See Table 5 of BS5589: 1989.
– indicates that the species cannot be preserved to satisfy BS5589: 1989 to the performance category by the organic solvent process indicated. It would be possible, however, to satisfy the performance category of BS5589: 1989 by a CCA process.

Not necessarily in the section dealing with the quality of timber, the specification should also include details of:

- *The paint or stain to be used.* It is essential that any paint or stain used on exterior surfaces is of exterior quality.
- *The adhesive type to be used* (or reference to the durability requirements in BS1186: Parts 1 and 2), and any special requirements such as weatherseals, compatibility of preservation/weatherseals/finish, compatibility of any filler with the finish, glazing method to be used etc.

Guidance on 'Joinery Grading' is given in Chapter B4, on 'Durability and Preservation' in Chapter B5, and on 'Painting' in Chapter B7.

SPECIES AND CLASSES

General

Of the species covered by this book, European Redwood from Sweden, Finland and Russia, European Whitewood from Sweden, Finland and Russia, and Canadian Douglas Fir-Larch have been used in the UK in significant quantities for timber windows. European Redwood is by far the most common. Canadian Hemlock is used to a certain extent. One would not expect to use Parana Pine, Spruce-Pine-Fir, Maritime Pine or British-grown Sitka Spruce for UK windows, these species not being considered to be suitable for window joinery, although Parana Pine is used for window boards. Canadian Western Red Cedar is used but normally only in a limited range of buildings such as those having Cedar cladding.

European Redwood

European Redwood is by far the most common timber used for windows produced in the UK. It machines well and easily. For windows, BS1186: Part 1: 1986 requires it to be preserved. It is easy to preserve and, for some cases, can be preserved to adequate levels by immersion. Normally, for windows, it is preserved with an organic solvent by a double vacuum process or by immersion. It can be preserved by an organic solvent double vacuum process to the 60 year (or 30 year) performance category of BS5589: 1989 or, by immersion, to the 30 year performance category.

The most relevant Classes of BS1186: Part 1: 1986 for European Redwood are Classes 2 and 3. It is possible for window sections of European Redwood to match Class 1 but only if the manufacturer is given time and the price to purchase a very high grade, or to reject a high percentage from more readily available grades, or to laminate. It is only possible to select European Redwood to match Class CSH for small, short pieces such as glazing beads.

If a stain finish is to be used, care must be taken to ensure that any noticeable discoloured sapwood ('blue stain') is excluded on Exposed and Semi-Concealed surfaces. Occasionally surface areas of high-porosity may be revealed, particularly if a dark stain is applied, by areas of deeper colour due to an excess of pigment being absorbed. Timber having areas of high-porosity should be avoided as much as is practical.

European Whitewood

Although European Whitewood has been used for many years in the UK for interior joinery and wood trim, it is only since the mid-1980s when guidance became generally available on how to preserve the species commercially, and on how to mould it to an acceptable finish, that it has been used to any significant extent for windows.

It is 'resistant' to preservation but, for windows, can be preserved commercially to the 30 year performance category of BS5589: 1989, normally in the UK by an organic solvent double vacuum process. It cannot be preserved adequately by immersion.

Largely because of the hard knots, care must be taken during moulding to ensure that the cutters are kept sharp and are set at the correct angle, and that the moisture content of the timber is carefully controlled. Even then, the surface of some knots may crack. If machining European Whitewood for joinery for the first time, specialist guidance should be sought.

The best qualities of European Whitewood for joinery come from particular areas of Sweden and Finland. Specialist timber buyers should be aware of these areas, but there is no need for reference to be made to them in a window specification intended for end-users.

The most relevant Classes of BS1186: Part 1: 1986 are Classes 2 and 3. It is possible for window sections of European Whitewood to match Class 1 but only if the manufacturer is given time and the price to purchase a very high grade, or to reject a high percentage from more readily available grades, or to laminate. It is only possible to select European Whitewood to match Class CSH for small, short pieces such as glazing beads.

Douglas Fir-Larch

Douglas Fir-Larch is usually only used for the production of more expensive timber windows. Although the species contains large knots, these are well-dispersed with useful lengths of largely knot-free timber between them, and manufacturers can purchase 'Clears' (see Chapter C5). It is normal to use Douglas Fir-Larch to match Classes 1 or CSH. It is not usual to call for the species to be preserved, the assumption in BS1186: Part 1: 1986 being that sapwood is excluded. If, however, the specifier does call for Douglas Fir-Larch to be preserved, it has to be treated as a species 'resistant' to preservation. Strictly speaking, therefore, although it is not normally considered necessary to preserve Douglas Fir-Larch for windows (even for a 60 year desired service life), if it is preserved it cannot be claimed to be preserved to the 60 year performance category (see Table 5 of BS5589: 1989) which seems to be somewhat of a contradiction.

Douglas Fir-Larch has prominent growth rings, often widely spaced, which may even be visible through an opaque paint. BS1186: Part 1: 1986 points to it as being susceptible to iron staining and recommends the use of non-ferrous fasteners in damp conditions.

Canadian Western Hemlock

BS1186: Part 1: 1986 rates Canadian Western Hemlock as being 'suitable' for windows if preserved. In future, it is possible that it may be used as an alternative to Douglas Fir-Larch, but this has not happened yet in the UK to any great extent. If, in future, sizes and lengths are produced in grades suitable for window joinery, this could result in its greater use.

For windows, BS1186: Part 1: 1986 requires Hemlock to be preserved. Although BS5589: 1989 permits Hemlock to be considered as being 'moderately resistant' when selecting a suitable preservation cycle for exterior door construction, for windows it requires Hemlock to be considered as being 'resistant'. For windows it can be preserved commercially to the 30 year performance category, usually in the UK by an organic solvent double vacuum process. For windows it cannot be preserved adequately by immersion.

ADHESIVES

In the assembly of windows, it is normal to use a urea or a catalysed or non-catalysed polyvinyl acetate adhesive (see Chapter B9). It is not normal to use a WBP adhesive.

If a polyvinyl acetate type adhesive is to be used and the window is to be finished with an opaque paint other than a very dark paint, an adhesive matching the B3 test requirements of German DIN Standard 68 602: 1979 on the strength of bond is normally adequate, whereas if a non-opaque finish or a very dark opaque finish is to be used, it may be necessary to use an adhesive matching the B4 test requirements of the standard (see the notes on DIN 68 602: 1979 in Chapter B9, and Tables B9.4 and B9.5).

If a urea or other resin adhesive is to be used, see Table B9.1 and the text in Chapter B9.

BS644: PART 1: 1989 *WOOD WINDOWS. SPECIFICATION FOR FACTORY ASSEMBLED WINDOWS OF VARIOUS TYPES*

There are cases where a specifier might be content with specifying that the timber in windows should comply with BS644: Part 1: 1989 *Wood windows. Specification for factory assembled windows of various types.* However, it is prudent to regard this standard as detailing minimum acceptable levels and, in some clauses, as giving guidance rather than full details. For example, the Class of timber specified for main frames is given as Class 3 of BS1186: Part 1: 1986, with Class 2 being required for 'casements, sashes and beads, except for small section beads which must be Class CSH'. For preservation, reference is made to BS5589: 1989 but not to either the 30 or 60 year desired service life performance categories.

TYPICAL SPECIFICATIONS

Each manufacturer or specifier of timber windows must make the final

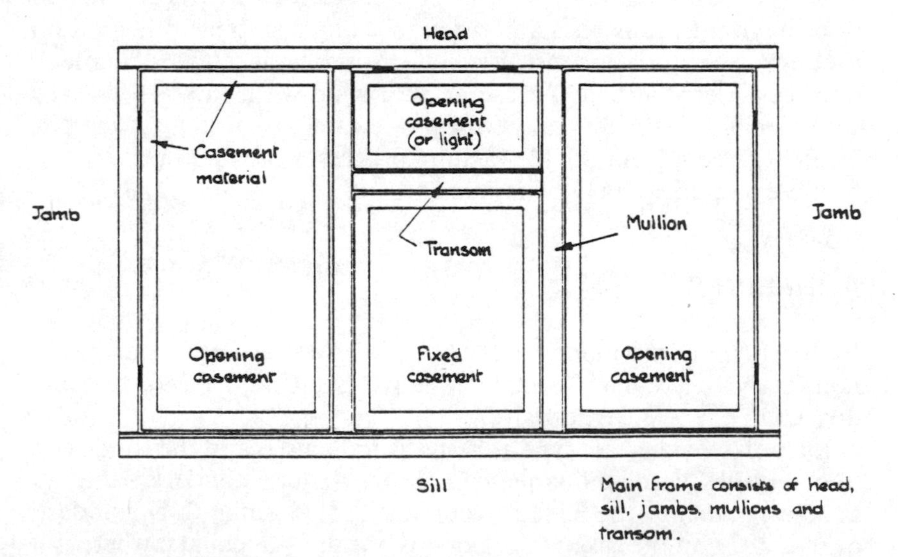

Fig. A1.1 Elevation on a typical timber window configuration to indicate terms used in the text

decision on what and how to specify, but the following are given as five examples of the general form of typical specifications for timber in windows. They do not cover all possible cases.

In the examples, 'casement material' is used to mean the timber in the opening frames (casements) or in a sub-frame fixed in the main frame even if not hinged to open (see Fig. A1.1).

Typical specification for timber in 'normal quality' Redwood windows to receive an opaque finish.

Windows to be finished with an opaque paint.

The timber is to be European Redwood. The frames (main) are to comply with Class 3 of BS1186: Part 1: 1986, the casement material with Class 2.

Moisture content of the timber at the time of manufacture and delivery to site is to be an average of 16% ± 3%.

The timber is to be preserved to the 30 year desired service life performance category level of BS5589: 1989 by an organic solvent double vacuum process or by immersion.

Workmanship is to be in accordance with BS1186: Part 2: 1987.

If a polyvinyl acetate adhesive is used it is to comply with the B3 requirements of DIN Standard 68 602: 1979 unless the manufacturer is informed in writing that a dark coloured paint is to be used, in which case it is to comply with the B4 requirements. If a resin adhesive is used it is to comply with the BR requirements of BS1204: Part 1: 1979 or BS1204: Part 2: 1979. In addition, a resin adhesive complying with the MR requirements of BS1204: Part 1: 1979 or BS1204: Part 2: 1979 may be used if the adhesive manufacturer provides evidence of its suitability for use even if a dark opaque paint is to be used.

Typical specification for timber in 'high quality' Redwood windows to receive a decorative stain (non-opaque) finish on exterior and interior surfaces.

The timber is to be European Redwood complying with Class 2 of BS1186: Part 1: 1986 with the following additional refinements:

Laminating of sections is permitted. Finger jointing of individual laminates is permitted but not of whole sections.

Splits, shakes, checks and plugs are to comply with the requirements of Class 1 of BS1186: Part 1: 1986. Inserts and fillers are not permitted. No stained sapwood or sign of insect attack is permitted on Exposed or Semi-Concealed surfaces.

Timber glazing beads to be European Redwood complying with Class CSH of BS1186: Part 1: 1986.

Moisture content of the timber at the time of manufacture and delivery to site to be an average of 16% ± 3%.

Timber to be preserved to the 30 year desired service life performance category level of BS5589: 1989 by an organic solvent double vacuum process.

If a polyvinyl acetate adhesive is used it is to comply with the B4 requirements of DIN Standard 68 602: 1979. If a resin adhesive is used it is to comply with at least the BR requirements of BS1204: Part 1: 1979 or Part 2: 1979.

Workmanship to be in accordance with BS1186: Part 2: 1987.

Typical specification by a manufacturer of timber in 'normal quality' softwood windows sold with a 'dual-purpose' primer for an opaque or non-opaque finish.

Timber in windows is European Redwood or European Whitewood complying with Class 3 of BS1186: Part 1: 1986 for the main frames and Class 2 for the material in the casements. Moisture content at the time of handover to the buyer is an average of 16% ± 3%.

Timber is preserved with an organic solvent to the 30 year desired service life performance category level of BS5589: 1989.

The adhesive is a polyvinyl acetate complying with the B4 requirements of DIN Standard 68 602: 1979.

Workmanship is in accordance with BS1186: Part 2: 1987. Where used, plugs, inserts, fillers and laminating comply with BS1186: Parts 1 and 2. Whole sections are not finger jointed.

Typical specification for timber in 'normal quality' softwood windows without the species being defined.

Timber in windows to be finished with an opaque paint is to be a softwood rated as 'Suitable' or 'Suitable if Preserved' in BS1186: Part 1: 1986 for windows, complying with Class 3 of BS1186: Part 1: 1986 for the main frames and Class 2 for the casement material. Different species may be used for the main frames and casement material (e.g. European Redwood for the frames and European Whitewood for the casements) but the use of species must be consistent for all windows, and species must not be mixed within one main frame or within one casement.

Moisture content at the time of manufacture and delivery to site is to be an average of 16% ± 3%.

Where BS1186 requires a species to be preserved, it is to be preserved with an organic solvent to the 30 year desired service life performance category of BS5589: 1989.

Workmanship is to be in accordance with BS1186: Part 2: 1987.

Adhesives are to comply with the requirements of BS1186: Parts 1 and 2.

Typical specification for timber in 'high quality' Douglas Fir-Larch windows to receive a decorative stain finish.

Windows to be finished with a decorative stain on exterior and interior surfaces.

Timber in main frames and casement material to be Douglas Fir-Larch complying with Class 1 of BS1186: Part 1: 1986 with timber glazing beads complying with Class CSH.

Laminating of sections is permitted. Finger jointing of individual laminates is permitted but not of whole sections. Inserts or fillers are not permitted. There should be no sign of insect attack on 'Exposed' or 'Semi-Concealed' surfaces. Moisture content of the timber at the time of manufacture and delivery to site is to be an average of 16% ± 3%.

If a polyvinyl acetate adhesive is used it is to comply with the B4 requirements of DIN Standard 68 602: 1979. If a resin adhesive is used it is to comply with at least the BR requirements of BS1204: Part 1: 1979 or Part 2: 1979.

Workmanship is to be in accordance with BS1186: Part 2: 1987.

Preservation is not required providing that sapwood is excluded.

TIMBER FOR DOORS, DOOR FRAMES AND LININGS A2

INTRODUCTION

In the majority of cases, a specifier or builder will choose a door or doors from a standard range offered by a manufacturer and will not prepare a detailed specification, but only refer to the door by a code generally accepted by specialists (e.g. 2XG, Pattern SA, Pattern 10, Pattern 50 etc., see Fig. A2.1), by a name given by the manufacturer, or by a description (e.g. framed ledged and braced, or ledged and braced, see Fig. A2.1, or framed and ledged). Even when a manufacturer prepares a door catalogue, it is quite usual for the details to be somewhat imprecise (e.g. to refer to BS1186: Part 1: 1986 on the quality of timber in joinery without referring to the Class of timber; to refer to 'Canadian Door Stock' without referring to the Class of timber or stating if Douglas Fir-Larch or Hemlock will be used). Manufacturers will, however, offer further details if asked and/or show a sample door. As an alternative, framed ledged and braced, ledged and braced, or framed and ledged doors can be specified to BS459: 1988 which covers 'matchboarded wooden door leaves for external use'.

EXTERIOR DOORS AND DOOR FRAMES

With the exception of 1 hour fire doors, exterior doors in the UK normally have a finished thickness of 44 mm. They can be designed as a flush door or a panel door. Typical panel doors such as Patterns 2XG, 10, 50 etc. are illustrated in Fig. A2.1. With panel doors, the stiles are usually machined from a basic width of 100 mm (see Chapter B1) or a 'green' sawn width of 4 inch (see Chapter C5), and are 44 mm thick. Some bow due to hygrothermal effects may occur in the height of the 'lock-stile' in certain weather conditions, particularly in winter if the temperature inside the building is high.

Exterior doors are subjected to varying service conditions and are largely unrestrained in position, therefore are free to 'move' in service. Experience has shown that it is usually unwise to use very wide stiles

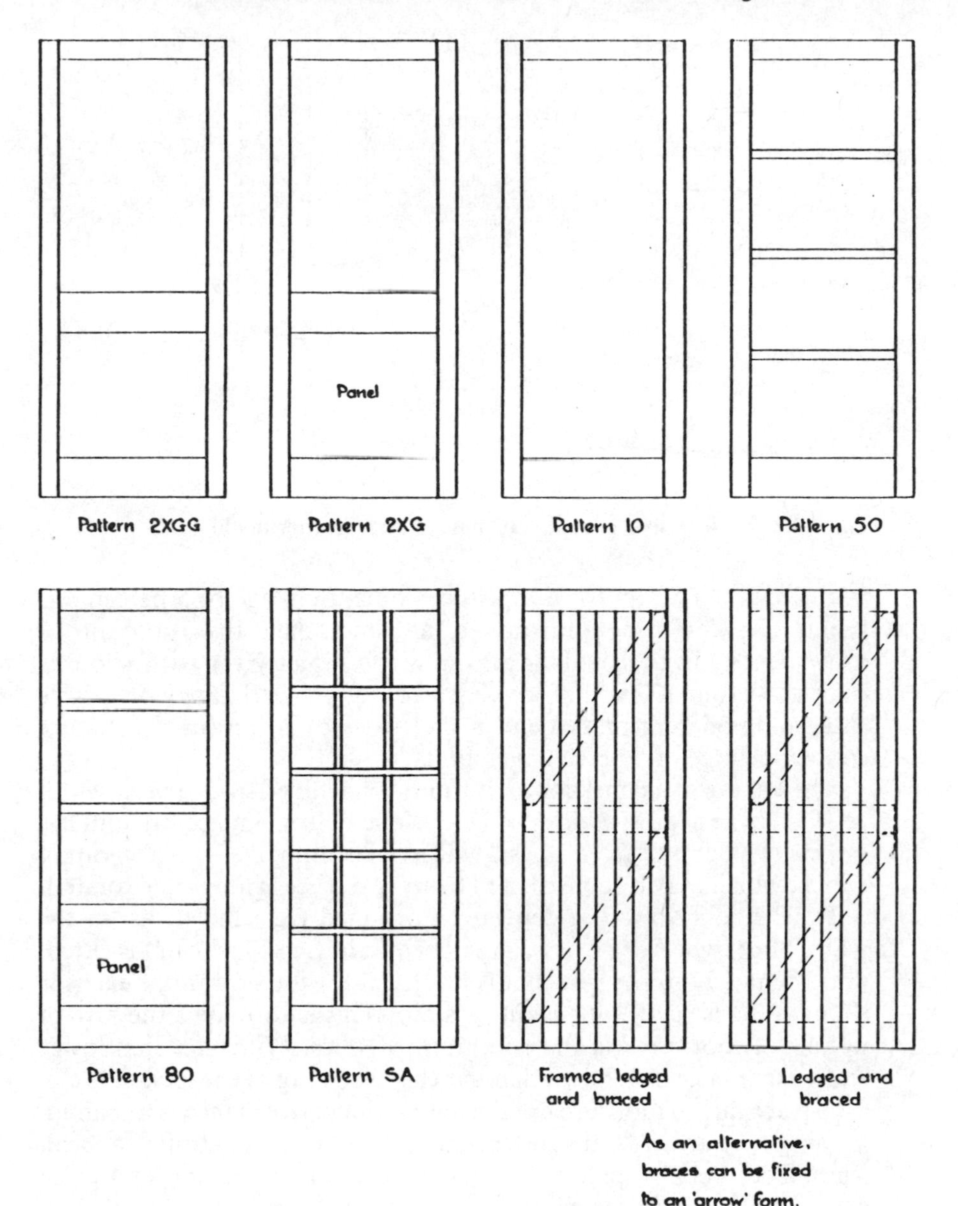

Fig. A2.1 Elevation of panel doors to illustrate pattern codes, and ledged and braced doors

(say out of 200 mm wide pieces). Any benefit which might accrue due to larger timber being used is likely to be off-set by the expansion/ shrinkage which can occur across the accumulated width of the stiles. For example, with a double door using four vertical stiles, each with a basic width of 200 mm, the total movement across the stiles would be

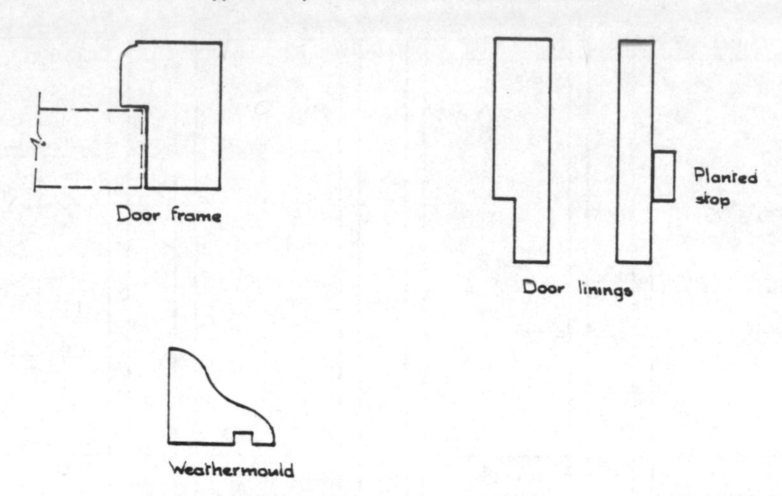

Fig. A2.2 Typical door frame, door linings and weathermould

in the order of 8 mm for a moisture content change of 5 percentage points. It would be difficult to accommodate that amount of movement in the design of the central vertical joint. Even with a double door using four 100 mm wide stiles, a central rebated joint would have to accommodate a movement in the order of 4 mm for a moisture content change of 5 percentage points.

The top rail of a panel door is normally machined from a basic width of 100 mm or a 'green' width of 4 inch. The bottom and centre rails are normally machined from a basic width of 200 mm or a 'green' width of 8 inch, which may be constructed from two pieces glued edge-jointed.

Panel doors are often delivered unprimed (referred to as 'in the white') but, whether or not a weathermould (see Fig. A2.2) is fitted, door manufacturers insist that it is critical for the bottom of exterior doors to be 'sealed' with paint or stain. This is to reduce the rate of uptake of moisture via the end grain of stiles. With unsealed lower ends, experience has shown that increased bowing of the lock-stile can take place due to moisture uptake, and paint performance is impaired. A single coat of stain or primer does not constitute a seal. Manufacturers also prefer the tops of exterior doors, and end grain exposed by mortises which go right through the door to be sealed.

An exterior door is fitted into a door frame which is usually moulded 'from the solid', although the rebate on the door frame can be formed by a piece 'glued and planted' into place.

INTERIOR DOORS

With the exception of fire doors, interior doors in the UK usually have an overall finished thickness of 35 mm or 40 mm.

Interior (non-fire) doors with an outside facing of hardboard, plywood etc. may have a softwood frame, a softwood block at the handle position, and a honeycomb core of paper or cardboard, with the outside facings glued to the frame and core. In such designs, the timber is normally dried to an average of 10% or 12% and is required to be free from cup and distortion, but the quality of the concealed surfaces is not particularly important and is rarely specified.

If an interior door is a panel door, it is normal for it to have the same appearance and width of members as an exterior panel door of the same style, but the finished thickness will be 35 mm or 40 mm. The infill panels can be glass, solid timber panelling, plywood etc.

An interior door is fitted into a 'door lining' or 'door casing' (see Fig. A2.2). Initially the term 'door lining' was used to describe a section machined 'from the solid', whilst a 'door casing' was a rectangular piece with a 'planted stop'. Nowadays it is more normal to use a planted stop on any frame around an interior door and for either term to be used to describe the frame. The normal practice is to deliver the stop 'loose' to enable the door to be 'hung' on the hinges before the stop is fixed although, if a pre-hung door set is delivered, the stop will normally be fixed in place.

FIRE DOORS

In certain situations dictated by Building Regulations, interior or exterior doors must be fire doors. To comply with the Building Regulations 1985 for England and Wales, they must satisfy the 'integrity' test requirements of BS476: Part 8: 1972 which covers *Test method and criteria for the fire resistance of elements of building construction*. To satisfy the Building Standards (Scotland) Regulations 1981 and the Building Regulations (Northern Ireland) 1977, they must satisfy the 'stability' and 'integrity' requirements of BS476: Part 8: 1972. The latest thinking is that fire doors should satisfy only the test requirements for 'integrity', and this is likely to be reflected in all future UK Regulations.

It is the intention that Part 8 of BS476 should be superseded by BS476: Part 22: 1987 on *Methods for determination of the fire resistance of non-loadbearing elements of construction* (and Part 20: 1989, Part 21: 1987 and Part 23: 1987), but Part 8 has not been withdrawn yet by BSI because it is still referenced in certain legislative documents.

Manufacturing members of the British Woodworking Federation (BWF) have adopted a coding of FD20, FD30 and FD60 for a 20, 30 or 60 minute resistance period respectively as quoted in BS5588: Section 1.1: 1984. The figure refers to the period in minutes of 'integrity' which sample doors have passed in tests to BS476: Part 8: 1972 (and/or Part 22). (If 'stability' is required, the specifier or purchaser should check if the design will satisfy the requirements. Note that there is no test for 'stability' in Part 22.)

For further reading, see Chapter B6 and the 'commentary' on fire doors in Clause 6.5.1. of BS5588: Section 1.1: 1984.

FD30 and FD60 doors are likely to require an intumescent strip to be positioned in the door or door frame, or door lining.

A standard FD20 or FD30 door is normally 44 mm thick whether of flush or panel construction. An FD60 flush door is normally 54 mm thick. No FD60 panel door is currently being made as standard. Fire doors of flush construction have a solid core.

Future Regulations may call for doors in certain positions to act as a barrier to smoke.

DOOR SETS

Dimensional requirements for door sets are contained in BS4787: Part 1: 1980.

SPECIFICATIONS

Specifiers are encouraged to choose doors, particularly fire doors, from standard ranges. If, however, a specifier insists on preparing or questioning a specification of a panel door, consideration can be given to the following Items (a) to (f). Flush doors are not considered here because of the great variety of facings which can be used.

(a) The species to be used

Popular species for more expensive panel doors are Canadian Douglas Fir-Larch or Canadian Hemlock. For cheaper panel doors, and ledged and braced, framed ledged and braced doors (see Fig. A2.1) and framed and ledged doors, European Redwood or European Whitewood are normally used.

Although Canadian timber can be used for door frames to match the species used in an exterior door, the species normally used are European Redwood or European Whitewood.

The species normally used for interior door linings are European Whitewood or European Redwood although the other species mentioned above can be used.

(b) The Class of timber

If Canadian Douglas Fir-Larch or Canadian Hemlock is used, normally Class CSH or Class 1 of BS1186: Part 1: 1986 is relevant for the quality of the stiles and rails in panel doors, and for door frames and linings if these species are used.

If European Redwood or European Whitewood is used for panel doors, with very careful selection it should be possible to match Class 1, although Class 2 is easier to match for cheaper doors intended to be finished with an opaque paint. The specifier may wish to consider tightening the clauses on shakes and checks.

For door frames and linings to be finished with an opaque paint, Class 2 would normally be considered to be adequate, although it may be desirable to specify Class 1 if a stain is to be used. When using European Redwood or European Whitewood for the stiles of panel doors or the longer members of door frames or linings, it is necessary to take care to select straight pieces.

For a ledged and braced, framed ledged and braced, or a framed and ledged door, to comply with BS459: 1988 the timber must be at least Class 3 of BS1186: Part 1: 1986, although the manufacturer or specifier may wish to tighten some clauses if a decorative stain is to be used. Standard boarded doors are not normally suitable to be finished with a decorative stain.

(c) The moisture content at the time of manufacture and delivery

The moisture content of a door, door frame or door lining should be appropriate for the service conditions as given in BS1186: Part 1: 1986 (see Table B2.2). As can be seen, these are an average of 16% for exterior joinery, and an average varying between 10% and 15% for interior joinery, therefore it is important for the specifier or purchaser to quote the service condition to the manufacturer. Simply to state 'to comply with BS1186' is not sufficient information.

(d) The sizes of members, tolerances and quality of machining

For panel doors, normally the basic sizes of members and the finished thickness will be as detailed in the introduction to this chapter. As far as the quality of machining and workmanship is concerned, it should be adequate to refer to BS1186: Part 2: 1987. For tolerances on timber sizes, refer to BS4471: 1987 (i.e. no minus tolerance permitted on planed dimensions, measured at 20% moisture content).

For a ledged and braced, framed ledged and braced, or framed and ledged door, to comply with BS459: 1988 the tongued and grooved matchboarding must be at least 15 mm thick. The dimensions and tolerances of the rails, stiles, ledges, braces and matchboarding are detailed in BS459: 1988. Note that a door with 'arrow headed braces' (i.e. with the upper and lower brace converging at one position at the middle rail or ledge) can be hinged on either stile, whereas doors with the braces parallel to each other should have the bottom of the lower brace at the hinge stile.

(e) Preservation, if required

If preservation of doors is considered to be necessary, in the UK it is normally carried out with an organic solvent type preservative. Therefore, in the notes below, CCA water borne preservatives are not considered. The relevant British Standard code is BS5589: 1989. For exterior joinery such as exterior doors, Table 5 of BS5589: 1989 details two desired service life performance categories which are 30 years and 60 years (see Chapter B5). For most normal building, if preservation of exterior doors and door frames is considered to be necessary, preserving to the 30 year performance category is normally considered to be adequate.

Preservation is not normally required for interior doors.

It is not normally considered necessary to preserve exterior panel doors of Douglas Fir-Larch, the assumption being that no sapwood is present.

If Canadian Hemlock is used, although it is normally rated as being 'resistant' to preservation, BS5589: 1989 permits it to be rated as 'moderately resistant' when it is being preserved for exterior doors by a double vacuum organic solvent process; see Table B5.3 for an explanation. This 'waiver' permits Canadian Hemlock to be preserved by a double vacuum organic solvent process (but not by immersion) to the 60 year as well as the 30 year performance category of BS5589: 1989. If, however, Hemlock is used for the frames around exterior doors, it should be considered to be 'resistant' to preservation, and consequently can be treated by a double vacuum organic solvent process only to the 30 year category (which should normally be considered to be adequate).

If European Redwood is used for exterior doors or exterior door frames, because it is 'moderately resistant' to preservation, it can be treated to the 30 year performance category of BS5589: 1989 by a 3 minute immersion in organic solvent, or by a double vacuum process, or by a double vacuum process to the 60 year category of BS5589: 1989. If applicable, pieces not thicker than 20 mm can be treated by a 60 minute immersion to the 60 year category.

If European Whitewood is used for exterior doors or exterior door frames, because it is 'resistant' to preservation, it can be treated by a double vacuum process (but not by immersion) to the 30 year

Table A2.1 Organic solvent preservation processes which satisfy BS5589: 1989 for exterior doors and door frames

Species and amenability of the heartwood to preservation[1]	Desired service life performance category and required organic solvent preservation process			
	30 years		60 years	
	Double vacuum	Immersion	Double vacuum	Immersion
Canadian Douglas Fir-Larch 'R'	Not normally required to be preserved			
Canadian Hemlock				
for doors 'MR' (see text)	+	–	+	–
for door frames 'R' (see text)	+	–	–	–
European Redwood				
not thicker than 20 mm 'MR'	+	+ (3 min)	+	+ (60 min)
thicker than 20 mm 'MR'	+	+ (3 min)	+	–
European Whitewood 'R'	+	–	–	–

[1] i.e. Resistant 'R' or Moderately Resistant 'MR' (see Table B5.3).
+ indicates that the species can be preserved to satisfy BS5589: 1989 to the performance category by the organic solvent process indicated. See Table 5 of BS5589: 1989 for the actual schedules.
– indicates that the species cannot be preserved to satisfy BS5589: 1989 to the performance category by the organic solvent process indicated. It would be possible, however, to satisfy the performance category of BS5589: 1989 by a CCA process.

performance category but not the 60 year category of BS5589: 1989.
This information is presented in Table A2.1.

(f) Fire resistance, if required

If a door (and door frame) is required to have fire resistance, it is vital for this information to be given to the supplier, and for the supplier of the door and door frame to be informed if the requirement is for an FD20, FD30 or FD60 door, and if the door must satisfy a requirement for 'stability' as well as 'integrity'.

TYPICAL SPECIFICATIONS

Each specifier or manufacturer of doors must make the final decision on what and how to specify, but the following are given as three examples of the typical content of specifications for doors and door frames.

Typical specification for exterior panel doors of Hemlock, with European Redwood door frames, all to be finished with an opaque paint.
Exterior panel doors to be 2XG, 1981 × 838 mm, 44 mm thick, Canadian Hemlock stiles and rails to Class 1 of BS1186: Part 1: 1986. Beads supplied loose to be Canadian Hemlock to Class CSH. Weathermoulds supplied loose to be Canadian Hemlock or European Redwood to Class 2. Door frames as Drawing ____ to be European Redwood to Class 2, with a hardwood sill to Class CSH of BS1186: Part 1 and a durability rating 'Suitable' or 'Suitable if Preserved' to satisfy Appendix C of BS1186: Part 1. All preserved timber to be preserved by an organic solvent to the 30 year desired service life performance category of BS5589: 1989. At the time of manufacture and delivery, average moisture content to be within the range 13–19% for the softwood and a maximum of 22% for the hardwood sill. Workmanship to be in accordance with BS1186: Part 2: 1987 with timber tolerances in accordance with BS4471: 1987.

Typical specification for interior panel doors of Hemlock to be stained, and European Whitewood door linings to be finished with an opaque paint.
Interior doors to be Pattern 50, 1981 × 762 mm, 35 mm thick, Canadian Hemlock to Class CSH of BS1186: Part 1: 1986. Beads supplied loose to be Canadian Hemlock to Class CSH. Door linings to be used with loose stops to be as Drawing ____ European Whitewood to Class 2 of BS1186: Part 1. Loose stops to be as Drawing ____, European Whitewood to Class 1 of BS1186: Part 1. At the time of manufacture and delivery, average moisture content to be within the range 10–14%. Workmanship to be in accordance with BS1186: Part 2: 1987 with timber tolerances in accordance with BS4471: 1987.

Typical specification for framed ledged and braced doors of European Redwood in door frames of European Redwood.
Exterior framed ledged and braced doors, and door frames to be as Drawing ____ and BS459: 1988 in European Redwood to Class 3 of BS1186: Part 1: 1986, with workmanship in accordance with BS1186: Part 2: 1987. At the time of manufacture, the average moisture content is to be between 13% and 19%. The doors and frames are to be preserved by an organic solvent process to the 30 year category of BS5589: 1989.

TIMBER FOR DOMESTIC STAIRS

A3

INTRODUCTION

BS5395: Part 1: 1977 (1984) is the code of practice for the design of straight stairs having an overall width not greater than 1220 mm, and with a total 'going' of any flight not exceeding 3800 mm. For stairs manufactured in accordance with BS5395: Part 1: 1977 (1984), the

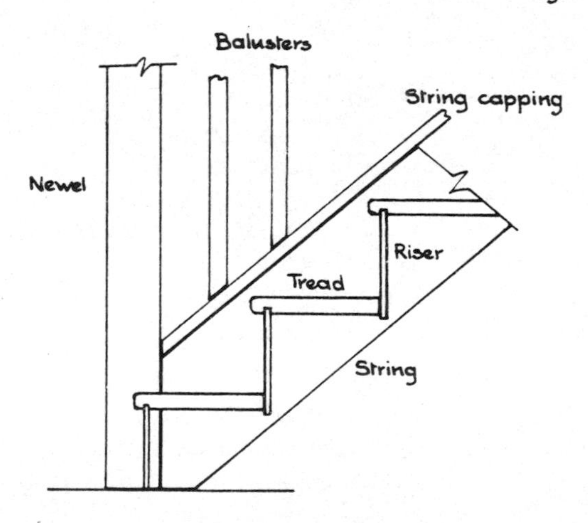

Fig. A3.1 Part elevation on a domestic stair to illustrate terms used in the text

required Class of timber (see below), sizes and tolerances of the strings, treads, risers, newels, handrails, balustrades, balusters and string cappings (see Fig. A3.1) are detailed in BS585: Part 1: 1989. BS585: Part 1: 1989 covers 'stairs with closed risers for domestic use, including straight and window flights and quarter or half landings'. BS585: Part 2: 1985 covers the 'performance requirements for domestic stairs constructed of wood-based materials'. Details of the Class of timber are given in BS1186: Part 1: 1986 on the quality of timber in joinery.

It is more usual for domestic stairs which are required to match BS585: Part 1 to be designed by manufacturers who offer standard ranges, rather than for specifiers to prepare a detailed specification for a manufacturer to follow. It is possible for a specifier simply to call for the material in stairs to comply with BS585: Part 1: 1989 and/or BS585: Part 2: 1985. However, a statement may be required about the type of finish which will be used (i.e. non-opaque or opaque) and, particularly if a non-opaque finish is to be used, it makes sense to specify the species which is (or are) to be used. Most standard stairs are supplied 'in the white' (i.e. unpainted) on the basis that the purchaser will decide what finish is to be applied. This is almost invariably the case for stairs supplied via non-manufacturing stockists. The following Items (a) to (d) give notes on the most likely options open to a specifier or manufacturer.

(a) The species to be used

The choice of species for the various parts of a stair assembly is partly dictated by the sizes of timber available in the various species. See Item (d). Although edge jointing and laminating can eliminate any restriction brought about by limits on the available sizes of sawn timber in a species, to date such processes have not been used to any great extent in the UK in the manufacture of stairs, although some strings are made in two pieces glued together edge-to-edge to give the necessary width.

Strings

The most popular species for the strings of domestic stairs is Brazilian Parana Pine. It is available in wide widths and is a fine looking timber with interesting colour variations, either knot free or with small pin knots (sometimes in considerable numbers). It is suitable for decorating with a clear finish although a non-opaque finish can be used. It has become rather expensive and European Whitewood is often used when an opaque finish is to be applied. BS1186: Part 1: 1986 on the quality of timber in joinery also lists Canadian Douglas Fir-Larch, Canadian

Hemlock, and European Redwood as being 'suitable' for stairs and these species are used to a certain extent.

Treads

If the strings are Parana Pine, it is quite normal (although not necessarily always the case) for the treads to be Parana Pine. European Whitewood or European Redwood, or Canadian Douglas Fir-Larch or Hemlock are also used. If the treads are to be left exposed, consideration should be given to the treads matching the strings (and the risers).

Risers

In modern stairs from standard ranges, it is usual for the risers to be made from a sheet material such as plywood rather than from solid timber. It may be possible to use a board with a face veneer which matches the strings.

Newels

Most newels are Canadian Douglas Fir-Larch or Canadian Hemlock. Parana Pine, European Redwood or European Whitewood can be used if the required size is available (or if laminated).

Handrails

European Redwood is easy to mould and is popular for handrails, as is Canadian Hemlock. Canadian Douglas Fir-Larch and Brazilian Parana Pine are also used.

Balustrades, balusters and string cappings

Canadian Hemlock is popular for balustrades and balusters, particularly if the newels and handrails are Hemlock. Douglas Fir-Larch and Brazilian Parana Pine are also popular, with European Redwood being less popular. European Whitewood can be used if the sections are mainly rectangular or square, but is not likely to be used if the balusters are turned or moulded (due to the difficulty in turning or moulding this species).

(b) The Classes of timber in BS1186: Part 1: 1986

With an Opaque Paint

BS585: Part 1: 1989 calls for the quality of timber in stairs which are to be finished with an opaque paint to comply with Class 3 of BS1186: Part 1 (see Chapter B4) except that handrails are to comply with Class 1.

With a Non-Opaque Paint

When a clear or non-opaque finish is to be used, BS585: Part 1: 1989 recommends that reference should be made to the manufacturer.

If Parana Pine, Canadian Hemlock or Douglas Fir-Larch is used, it should be possible to match Class CSH or Class 1.

If European Whitewood or European Redwood is used, it should be possible to match Class 2, and perhaps even Class 1 if laminating or edge jointing is used. The specifier should, however, check on the suitability of the species for the finish to be applied and for any 'making good' used by the manufacturer, particularly if a clear finish is to be used.

(c) The moisture content

The moisture content should comply with the recommendations of BS1186: Part 1: 1986 for interior joinery. The moisture content at the time of manufacture should be close to the in-service conditions as shown in Table A3.1 which are in accordance with BS1186: Part 1: 1986. Note that the moisture contents are average values, not ranges of permissible moisture contents.

(d) The sizes, tolerances and quality of machining

The quality of machining should comply with BS1186: Part 2: 1987. In

Table A3.1 Average in-service moisture contents

In-service condition	Average percentage moisture content in service
For buildings with intermittent heating	15 ± 2%
For buildings with continuous heating providing room temperatures of 12°C to 19°C	12 ± 2%
For buildings with continuous heating providing room temperatures of 20°C to 24°C	10 ± 2%

general, if planed rather than sanded, the number of cutter marks per 25 mm should be between 12 and 18.

BS585: Part 1: 1989 does not give specific guidance on the tolerances permitted on the size of members, other than to state that sizes which are given as minimum are finished minimum sizes. The dimensions given are measured as at 20% moisture content (as required in BS4471: 1987), and one presumes that the tolerances permitted in BS4471 for planed timber apply (i.e. – 0 mm + 1 mm).

Strings

To comply with BS585: Part 1: 1989, the finished thickness of strings should be not less than 26 mm, and the finished width of strings should be not less than 215 mm.

Normal basic sawn sizes (in mm) from which strings are planed are:

32 × 225	32 × 250	32 × 275	32 × 300
	38 × 250	38 × 275	38 × 300

If edge-to-edge jointing is used, no more than two pieces each greater than 50 mm are permitted. If a laminated section is used, the restriction on there being only two pieces does not apply. (For the purpose of BS585: Part 1: 1989, the definition of a laminated section, i.e. 'glue-laminated wood' is 'Wood consisting of laminations glued together, no constituent lamination being more than 50 mm or less than 8 mm thick, with the exception of facing laminations'.)

Treads

To comply with BS585: Part 1: 1989, the finished thickness of treads should be not less than:

- 20 mm for a width of stair (overall the strings) of up to 990 mm;
- 26 mm for a width of stair (overall the strings) exceeding 990 mm but not more than 1220 mm.

If edge-to-edge jointing is used, no piece of timber should be less than 50 mm wide on the finished face, and the front piece on which the nosing is formed should be not less than 90 mm wide. If a laminated section is used, these restrictions do not apply.

Risers

To comply with BS585: Part 1 : 1989, the finished thickness of risers should be not less than 14 mm if timber, and not less than 9 mm if plywood.

Newels

To comply with BS585: Part 1: 1989, outer newels (i.e. not fixed to a wall) should have a square cross-section not less than 69 × 69 mm, or another cross-section of equivalent strength and stiffness. Wall newels should have a cross-section of not less than 32 × 69 mm, or another cross-section of equivalent strength and stiffness.

Handrails, balustrades, balusters and string cappings

BS585: Part 1: 1989 calls for the dimensional designs of handrails and balustrades to be in accordance with BS5395: Part 1 and gives sketches of typical sections.

ADHESIVES

BS585: Part 1: 1989 permits the following adhesives to be used in the assembly of stairs. Workmanship should comply with BS1186: Part 2: 1987.

- Synthetic resin gap-filling adhesives (phenolic and aminoplastic) complying with the requirements of Type BR or MR of BS1204: Part 1: 1979 *Synthetic resin adhesives. Specification for gap-filling adhesives.*
- One part polyvinyl acetate emulsion adhesives complying with BS4071: 1966 *Specification for polyvinyl acetate emulsion adhesives for wood.*
- Two part polyvinyl acetate emulsion adhesives in accordance with the recommendations of EN 204, *Evaluation of non-structural adhesives for joining of wood and timber derived products* which is in the course of preparation.

Although not included in BS585: Part 1: 1989, animal glues in accordance with BS745: 1969 *Specification for animal glue for wood* are still used.

TYPICAL SPECIFICATIONS

Each manufacturer or specifier of timber in stairs must make the final decision on what and how to specify, but the following are given as two examples of the typical content of specifications for domestic stairs and associated timber.

Typical specification by a manufacturer, of the timber in stairs to BS585: Part 1: 1989, with Parana Pine strings, suitable for a clear finish.
The timber in strings and treads in the standard range of stairs shown in this catalogue is Brazilian Parana Pine matching Class CSH of BS1186: Part 1: 1986 with sizes and tolerances in accordance with the drawings and BS585: Part 1: 1989. Risers are plywood faced with ____. Workmanship is in accordance with BS1186: Part 2: 1987. The adhesive used in manufacture is a one-part polyvinyl acetate complying with BS4071: 1966. A range of newels, handrails, balustrades, balusters and string cappings is available in Canadian Hemlock or Canadian Douglas Fir-Larch. The average moisture content of the timber at time of manufacture and delivery from the factory is between 10% and 14%.

Typical specification by a manufacturer, of the timber in stairs to BS585: Part 1: 1989, with European Whitewood strings, suitable for an opaque finish.
The timber in strings and treads in the standard range of stairs shown in Drawing ____ is European Whitewood matching Class 3 of BS1186: Part 1: 1986, suitable for finishing with an opaque paint. The sizes and tolerances are in accordance with Drawing ____ and BS585: Part 1: 1989. Risers are plywood. Workmanship is in accordance with BS1186: Part 2: 1987. The adhesive used in assembly is a one-part polyvinyl acetate complying with BS4071: 1966. A range of handrails, balustrades, balusters and string cappings is available as shown in Drawing ____ in Canadian Hemlock or European Redwood. The average moisture content at time of manufacture is 12% ± 2%.

A4 TIMBER FOR INTERIOR WOOD TRIM

SUCH AS SKIRTING BOARDS AND T & G BOARDING

INTRODUCTION

Until recently there has been understandable confusion about the grading rules or British Standard to refer to when specifying the quality of timber in visible interior wood trim such as skirting boards and tongued and grooved boarding. One tendency was to refer to a sawmill grade such as 'fifth quality' or 'unsorted' when specifying Nordic timber but, as explained in Chapter B4, these grading rules are not precise, are not end use grades and are invalidated if re-sawing takes place. As an alternative, some specifiers have referred to one of the Classes of BS1186: Part 1, but this standard is intended for joinery and is over-complex for a specification for components such as skirting boards, interior tongued and grooved boarding etc. In fact, the 1986 version makes it clear that BS1186: Part 1 does not apply to wood trim.

The situation on specifying wood trim should be clarified by the publication, probably in 1990, of BS1186: Part 3 on *Wood trim: timber species, classification, workmanship in fixing, and standard profiles.* In the notes below, the assumption is made that BS1186: Part 3 will be published generally as drafted at the time of 'public comment' in 1988 and as cleared, in 1989, for publication by BSI. The specifier should examine the final document when available.

The point is made in the Foreword of BS1186: Part 3, that a considerable amount of wood trim is sold 'as seen', in which case there is no need to refer to the standard. The intention of BS1186: Part 3 is, however, that it will be available for the many occasions when a written specification is necessary. It is intended as a simplified form of BS1186: Part 1: 1986 on the quality of timber in joinery, but also includes a few clauses on fixing. It is designed to be used as the basis for specifications of exterior barge boards, fascia boards, architraves, weather trims, exterior timber cladding and ancillary trim (see Chapter A15), plus interior skirting boards, architraves etc. and interior decorative tongued and grooved boarding and ancillary trim. To coincide with the publication of BS1186: Part 3, the intention is that BS584: 1967 (1980) *Specification for wood trim (softwood)* will be withdrawn.

Particularly for interior trim, a range of standard sections is given in BS1186: Part 3, with standard codings to aid cross-referencing, particularly if a drawing is not available.

SPECIFICATIONS

A specification for timber in visible interior 'wood trim' as defined in the draft of BS1186: Part 3 as:

> skirting boards, picture rails, architraves, cornices, mouldings (such as scotia, half-rounds, astragals, cover-strips and fillets), interior solid timber panelling (tongued and grooved, square-edged, ship-lap), and ancillary trim, window boards etc.,

should include details of the following Items (a) to (d) which may vary in the form of the wording used depending whether it is written by a specifier or a supplier.

Note that items supplied 'loose' but intended to be fixed to joinery and to be a functional part of the joinery (e.g. handrails, newel posts, door weathermoulds etc.) are to comply with BS1186: Parts 1 and 2 rather than BS1186: Part 3.

(a) The species which is to be used or the list of species which are acceptable or could be used

The most common species for interior wood trim including interior solid timber panelling is European Redwood. It is readily available in a wide variety of sawn sizes and it is easy to plane and mould. European Whitewood is also used in quantity but, although it is as readily available as European Redwood, it is not as easy to plane or mould unless the manufacturer has the necessary equipment and expertise.

If available in suitable sizes, Parana Pine is very suitable for most interior trim, as is Canadian Hemlock, Canadian Douglas Fir-Larch and Canadian Western Red Cedar. Some interior wood trim is made from Maritime Pine.

(b) The Class of timber required in the final work as fixed

If European Redwood or European Whitewood is used, small pieces may be able to match Class 1 or even Class CSH of BS1186: Part 3. However, in the majority of cases, particularly for larger sections, Classes 2 or 3 are likely to be the most relevant. One of the most common and popular uses of European Redwood is decorative interior tongued and grooved boarding ('matchings'). It is worth noting that, for this specific case (and for exterior cladding), the latest draft of BS1186: Part 3 permits pith to occur on 'Exposed' surfaces of Classes 1, 2 and 3 (but not Class CSH).

If Parana Pine, Canadian Western Red Cedar, Canadian Hemlock or Canadian Douglas Fir-Larch is used, the most relevant Classes of BS1186: Part 3 are Class 1 and Class CSH.

Most wood trim is supplied 'loose' to site and has one or more ends trimmed before being fixed. BS1186: Part 3 takes account of this and, as drafted, permits 70 mm at every end as supplied to have defects or characteristics larger than the Class limits, on the basis that end trimming will take place at the time of fixing to bring the pieces within grade.

(c) The moisture content at the time of delivery and fixing

The latest draft of BS1186: Part 3 calls for 'interior wood trim' as defined to have a moisture content at the time of handover to the purchaser and before fixing between the limits shown in Table A4.1. It is desirable for the trim to be delivered and fixed at a moisture content close to that which will be encountered in service. As likely to be printed, BS1186: Part 3 gives a range of limits, not average values and a maximum as in Part 1: 1985.

It may well be fairly easy to obtain wood trim at a moisture content as low as about 14–16%, particularly if the sections are thin. However, if lower moisture contents are required, it will usually be necessary to place a special order, to pay extra, and to allow the supplier enough time to arrange further drying.

Table A4.1 Limits of moisture content of interior wood trim based on service conditions

Heating conditions and temperature	Limits of moisture content at time of fixing (%)
For buildings with intermittent heating	13–17
For buildings with continuous heating providing a room temperature of 12–19°C	10–14
For buildings with continuous heating providing a room temperature of 20–24°C	8–12

(d) Sizes, tolerances and surface quality

BS1186: Part 3 as drafted includes details of certain standard profiles of:

tongued and grooved boards,
skirting boards,
architraves,
half-rounds, cover-fillets, quadrants and scotia mouldings

which are intended to be useful to specifiers and designers. Standard codings are given to aid cross-referencing, particularly if a drawing is not available.

The tolerances of BS4471: 1987 on sizes of sawn and processed softwood apply (i.e. no minus tolerance permitted on processed sizes measured at 20% moisture content). See Chapter B1 for further relevant information on BS4471: 1987.

'Exposed' surfaces can be sawn, planed or sanded. The 'Exposed' surfaces of most visible interior wood trim used in the UK is planed or sanded. If planed, the surface should comply with BS1186: Part 3: 1986 (i.e. there should be between 12 and 18 cutter marks per 25 mm).

Note that the draft of BS1186: Part 3 includes clauses on the extent of making good which is permitted at the time of decoration.

SPECIAL TREATMENTS

It is not often that it will be necessary to preserve interior wood trim. If preservation is required in special cases see BS5589: 1989 on preservation of timber, and Chapter B5.

When interior timber boarding is specified, it may be necessary to consider any regulatory requirements for a 'surface spread of flame' rating. The requirements of Building Regulations are too detailed to include in this book although some background guidance is given in Chapter B6.

TYPICAL SPECIFICATIONS

Each builder or specifier of timber must make the final decision on what and how to specify, but the following are given as three examples of the typical content of specifications for some interior wood trim as defined in BS1186: Part 3, on the assumption that the final document will be published generally as drafted at the time of 'public comment' in 1988 and as cleared, in 1989, for publishing by BSI.

Typical specification for skirting boards.
Skirting boards are to be European Redwood of the profile and tolerances shown in Drawing ____(i.e. Standard profile No. ____of BS1186: Part 3: 19__), to comply with Class 2 and the machining requirements of BS1186: Part 3. The moisture content at time of delivery and fixing is not to exceed 14%. Boards are to be supplied in random lengths not less than 2.4 m long and are to be nailed into place, with the standard of fixing complying with BS1186: Part 3. At the time of decoration, 'Exposed' surfaces are to be made good to comply with BS1186: Part 3 suitable for an opaque gloss paint.

Typical specification for interior vertical tongued and grooved solid timber boarding, fixed to horizontal battens, and not required to meet any 'surface spread of flame' requirements.
The tongued and grooved boarding is to be European Redwood of the profile and tolerances shown in Drawing ____(i.e. Standard profile No ____ of BS1186: Part 3: 19__) supplied in basic lengths of 2.4 m and complying with Class 3 of BS1186: Part 3 with pith permitted on 'Exposed' surfaces. At the time of delivery, the moisture content is to be between 10% and 14%. 'Exposed' surfaces are to be planed in accordance with BS1186: Part 3.

Typical specification for timber in window boards.
Window boards are to be Parana Pine of the profile, tolerances and lengths shown in Drawing ____ complying with Class CSH of BS1186: Part 3. At the time of delivery and fixing, the moisture content is to be between 10% and 14%. 'Exposed' surfaces as defined in BS1186: Part 3 are to be planed in accordance with BS1186: Part 3. Edge jointing or laminating in accordance with BS1186: Part 2 is permitted, but finger jointing is not permitted.

A5

TIMBER FOR SEMI-STRUCTURAL USES

SUCH AS BRACING, RIDGEBOARDS, WALLPLATES, BATTENS FOR EXTERIOR CLADDING AND INTERIOR BOARDING, AND CONCEALED NON-STRUCTURAL FRAMING

INTRODUCTION

There have been significant improvements in recent years in the production of practical British Standards on stress grading (i.e. BS4978: 1988 on softwood grades for structural use), on the quality of timber in joinery (i.e. BS1186: Part 1: 1986, and Part 2: 1987 on workmanship), and wood trim including cladding, fascia and barge boards (i.e. BS1186: Part 3 when published probably in 1989 or 1990, including clauses on fixing and standard profiles). There are still, however, several unsophisticated uses of timber which are not covered by a British Standard (unless one is prepared to over-specify by calling-up one of the standards mentioned above). Some of the more common examples of these uses are:

- Diagonal and longitudinal roof bracing such as the bracing in a trussed rafter roof.
- Wallplates on top of a masonry wall. (Wallplates on top of a timber stud wall are covered in general by the same standards as the other timber in the stud wall.)
- Gable ladders as in a trussed rafter roof.
- Ridgeboards in a traditional 'cut' pitched roof.
- Solid blocking, noggings (non-loadbearing), or herringbone strutting between joists.

- Battens fixed to a wall to support exterior claddings (such as timber cladding or tile hanging), or battens supporting interior panelling.
- Concealed non-structural timber framing.

Although there will be several instances where the specification will include details of preservation requirements, often the only quality requirement of the timber for the cases outlined above is that it should be reasonably dry at the time of fixing, and it should be free from decay and live insect attack or any major defect which would make it likely to break whilst being fixed in place, or in service. In some cases, it should be free from wane. Normally it should be free from any significant distortion which would make it difficult to be fixed into place. Even with unsophisticated uses, however, if one wishes to avoid disputes on site as to what is or is not acceptable, the specifier may have no option but to write a more detailed specification than the component would seem to justify. In such a case, of course the quality specification should not be tighter than is required. One should not over-specify. Some examples are discussed below.

The specifier or builder has the option of:

(1) Writing a specification without cross-reference to any British Standard.

(2) Referring to a British Standard such as BS4978 on stress grading (see Chapter B3), BS1186: Part 1 on joinery (see Chapter B4), or BS1186: Part 3 (when published) on wood trim (see Chapter B4), but perhaps excluding or relaxing the need to comply with certain clauses to try to avoid over-specifying.

(3) Extracting (and perhaps relaxing) a few relevant clauses from a standard such as BS4978, BS1186: Part 1 or BS1186: Part 3 without actually referring to the standard.

(4) Agreeing quality from a sample (e.g. selling or buying timber 'as seen'), which is a point made in the Foreword of BS1186: Part 3 as a common method of purchasing wood trim.

One should not over-specify. It may not even be necessary to limit species but, as can be seen from reference to the notes and examples below, it is difficult to be both brief and precise.

(5) Additionally, one tendency has been to specify a sawmill grade such as 'fifth quality' for Nordic timber. However, as explained in Chapter B4, the grading rules for sawmill grades are not precise, some refer only to the 'better' face and two edges, and the original grading is invalidated if re-sawing takes place. Provided that this is realized, however, in cases where the timber is not to be resawn,

and where there is no need for a precise description of quality, and where the buyer is familiar with sawmill grading (or perhaps is shown a typical sample delivery), it may be adequate for the end use specification to be based on a sawmill grade.

BACKGROUND NOTES ON SPECIFICATIONS FOR A SELECTION OF UNSOPHISTICATED TIMBER USES

For the purpose of this chapter, a mixture of background notes and suggested typical specifications are given, with as little reference to British Standards as is sensible to clarify or simplify a specification. It must be emphasized, however, that the final decision on what and how to specify rests with the specifier, manufacturer or builder in each specific case.

With the exception of the details in Table A5.1 which cover preservation of wallplates on top of a masonry wall to satisfy BS5268: Part 5: 1989 on the preservative treatment of structural timber, preservation is not covered to any extent in the following notes. The reason is that preservation is covered in detail in other chapters, e.g. in Chapter B5 in general terms, and in several chapters dealing with specific situations, such as Chapter A7 on timber (such as bracing) in trussed rafter roofs. Cross-reference to these chapters is given in each case discussed below.

Diagonal and longitudinal bracing in a trussed rafter roof

The minimum size of diagonal and longitudinal standard bracing members for trussed rafters spaced at centres not exceeding 600 mm is given in BS5268: Part 3: 1985 as 22 × 97 mm. It is normal to use a basic sawn size of 22 × 100 mm European Whitewood with the tolerances of BS4471: 1987, but BS5268: Part 3: 1985 also permits European Redwood, Canadian Hem-Fir, Douglas Fir-Larch and Spruce-Pine-Fir, USA timbers (not covered by this book), and British-grown Scots Pine and Corsican Pine (note: not British-grown Sitka Spruce) to be used for standard bracing.

If specially designed bracing is used, it should be stress graded, but there is no requirement in BS5268: Part 3: 1985 for standard bracing to be stress graded. The only grading requirement in BS5268: Part 3: 1985 for standard bracing is that it should be 'free from major strength reducing defects', which is rather a vague specification for a British Standard. If, for a special contract, it is necessary to define 'free from major strength reducing defects', unless a stress grade must be specified it is probably adequate to call for:

Bracing members to be free from live insect attack, decay and any significant deformation which would require them to be forced into place; to be free from wane and splits at connection positions; to have a knot area ratio not in excess of one half; and for the slope of grain not to exceed 1 in 6. At the time of fixing, the moisture content should not exceed 24%.

If the trussed rafters or trusses are preserved, it is normal also to preserve the bracing for the same risk category and to the same level as required to satisfy BS5268: Part 5: 1989 (see Table A7.1).

Wallplates fixed on top of a masonry wall and a timber stud wall

In the case of 'wallplates' which are the top horizontal members of timber stud walls, it is normal for the specification of the 'wallplates' including any preservation requirements, to be generally the same as for the other timber in the stud wall (see Chapter A10).

Wallplates fixed to the top of a masonry wall normally have a basic width of at least 75 mm and a basic thickness of at least 47 mm (although sometimes as thin as 38 mm). A common size is 75 × 100 mm. The most common species is European Whitewood but many species can be used if timber of suitable size is available in that species. The tolerances would normally be expected to match BS4471: 1987 unless NLGA sizes and tolerances are specified (see Chapter C5).

There is no requirement for the timber to be stress graded.

Normally it should be adequate to specify that the timber should be free from wane, decay and live insect attack, and any significant deformation which would require it to be forced into place. Even if a more detailed specification is considered, it seems unnecessary to place a limit on splits, checks and shakes, rate of growth, slope of grain or the size of knots.

At the time of fixing, the moisture content should not exceed 24% (even if the timber has been treated with a water borne preservative).

Normally it will not be necessary for a specifier to detail the actual schedules which are permissible but, for information for the cases when it is decided to preserve wallplates, the schedules detailed in Table A5.1 satisfy Table 4 of BS5268: Part 5: 1989. Note that, for certain risk categories, the schedules required for wallplates are more onerous than those for other timbers in a roof. In an area where House Longhorn Beetles are active, it is normal to consider the wallplates to be within the roof void and to preserve them against Beetle attack. In an area where House Longhorn Beetles are not active, in a 'dry' pitched roof and a 'warm' flat roof, there is probably a 'low risk' or 'negligible risk' of fungal or insect attack (see Tables B5.4 and B5.5),

Table A5.1 Treatment schedules and immersion periods for wallplates in pitched or flat roofs in a building to satisfy BS5268: Part 5: 1989

	'Permeable' or 'moderately resistant' species 'P' or 'MR' [3]		'Resistant' or 'extremely resistant' species 'R' or 'ER'	
	Organic solvent	**CCA water borne**	**Organic solvent**	**CCA water borne**
In an area where House Longhorn Beetles are active (i.e. Risk 4 'Essential' for insect attack[1]) where there is considered to be 'negligible' risk of fungal attack or other insect attack[1]				
'Roof void' in a pitched or flat roof	Schedule V/1 or 10 minute immersion[2]	Schedule P3 20 g/l solution	Schedule V/1 or 10 minute immersion[2]	Schedule P3 30 g/l solution
In a situation where Furniture beetles are a 'low risk' (i.e. Risk 2 'Optional' for insect attack[1]) but where there is 'negligible risk' of fungal attack[1] but it is decided to preserve only against insect attack				
'Dry' pitched roof, Flat roof[5] 'cold', or Flat roof 'warm'[5]	Schedule V/1 or 3 minute immersion[2]	Schedule P2 20 g/l solution	Schedule V/1 or 10 minute immersion[2]	Schedule P2 30 g/l solution
In a situation where there is a 'low risk' of fungal attack[1] and 'Negligible risk' of insect attack[1] but it is decided to preserve against fungal attack				
'Dry' pitched roof	Schedule V/1 or 3 minute immersion	Schedule P2 20 g/l solution	Schedule V/3 or V/4[4]	Schedule P2 30 g/l solution
In a situation where there is a 'low risk' of fungal attack[1] and 'Negligible risk' of insect attack[1] but it is decided to preserve against fungal attack				
Flat roof 'cold'	Schedule V/2	Schedule P3 20 g/l solution	Schedule V/4	Schedule P7 30 g/l solution
Flat roof 'warm'	Schedule V/1	Schedule P3 20 g/l solution	Schedule V/3 or V/4	Schedule P7 30 g/l solution

For cases where there is considered to be a risk of both fungal and insect attack, use the more onerous of the schedules or immersion periods.

1 See Tables B5.4 and B5.5, and Tables 1, 2 and 3 of BS5268: Part 5: 1989 for further explanation and application of the 'Risk categories'.
2 The organic solvent must include an insecticide if the risk is insect attack.
3 When preserving mixed species, see the relevant paragraph of Chapter B5.
4 When treating 'Resistant' or 'Extremely resistant' species by an organic solvent against fungal attack, immersion is not recommended for wallplates.
5 Assuming that a wallplate for these flat roofs has the same risk as for a pitched roof.

and only slightly more risk of fungal attack in a 'cold' flat roof. In each case, the specifier will have to decide whether or not to preserve.

Gable ladders as in a trussed rafter roof

Gable ladders can be made from sawn timber normally of at least 38 mm basic thickness in which case European Whitewood or European Redwood are the species most commonly used, or from

38 mm thick surfaced CLS Canadian Spruce-Pine-Fir. In cases where gable ladders are pre-made by a manufacturer of trussed rafters, it is quite common for them to be made from the same material as the trussed rafters (e.g. processed to 35 × 72 mm or 35 × 97 mm) or, if the manufacturer assembles timber stud walls, from surfaced 38 × 89 mm CLS. Tolerances should be to BS4471: 1987. There is no requirement for the timber to be stress graded.

Normally it should be adequate to specify that the timber should be free from wane, decay and live insect attack, and any significant deformation which would require it to be forced into place, and to be free from any major strength-reducing features. Even if a more detailed specification is considered, it seems unnecessary to place a limit on splits, checks and shakes, rate of growth or slope of grain. It even seems unnecessary to place a limit on the 'knot area ratio' (see Chapter B3). Sometimes the material used for gable ladders is timber rejected at the time of stress grading, or 'short ends' of stress graded timber. In general, experience shows this material to be adequate.

At the time of assembly and fixing, the moisture content should not exceed 24%. If the trussed rafters are preserved, it is normal to preserve the gable ladders for the same risk category and to the same level as required to satisfy Table 4 of BS5268: Part 5: 1989 (see Table A7.1).

Ridgeboards in a traditional pitched 'cut' roof

Ridgeboards in a traditional pitched 'cut' roof are normally basic sawn sizes, probably 22 mm thick if of European Whitewood or 25 mm thick if of European Redwood. They could also be British-grown Sitka Spruce, Canadian Spruce-Pine-Fir or Hem-Fir. The depth is usually 150, 175, 200 mm or deeper depending on the size and slope of the incoming rafters. Tolerances should be to BS4471: 1987.

Normally it should be adequate to specify that the timber should be free from wane, decay and live insect attack, and any significant deformation which would require it to be forced into place. Even if a more detailed specification is considered, it seems unnecessary to limit the rate of growth, slope of grain, or even the 'knot area ratio' (see Chapter B3).

One purpose of using a ridgeboard is to provide a timber surface to which rafters can be nailed conveniently, therefore it is sensible to place a limit on splits, checks and shakes. Fairly short (say 50 mm long) end splits (i.e. through the piece) should be acceptable, with the length of individual shakes or checks (i.e. not right through the thickness) being limited say to 300 mm.

The moisture content at the time of fixing should not exceed 24%. If the other roof timbers such as rafters are preserved, it is normal to preserve the ridgeboards for the same risk category and to the same level as required to satisfy Table 4 of BS5268: Part 5: 1989 (see Table A8.1).

Solid blocking, noggings (non-loadbearing) or herringbone strutting between joists

Solid blocking, noggings (non-loadbearing) or herringbone strutting between joists can be a variety of sizes but usually has a basic thickness of not less than 38 mm. Although sawn timber is normally used, if solid blocking is used for the full depth of joists, the blocking will often be the same material as the joists (e.g. regularized to BS4471: 1987, or surfaced CLS) and will have the same tolerances. Sometimes it will be slightly less deep than the joists. If noggings or herringbone struttings are used, the tolerances should comply with BS4471: 1987. Normally any European or Canadian softwood species is acceptable. Particularly if the timber is preserved, there are advantages in using the same species as the joists.

If solid blocking is used, the timber should be free from decay and live insect attack and have the same limit on wane as the joists. It seems unnecessary to place a limit on the knot area ratio, rate of growth or slope of grain, but, to ensure solid end fixing to the joists, end splits should be limited at the time of fixing.

If small sections (say 47 × 50 mm basic sawn sizes) are used for non-loadbearing noggings, or small sections for herringbone strutting, the timber should be free from wane, decay and live insect attack, it is preferable to limit the knot area ratio (say to ½), but it hardly seems necessary to limit the rate of growth or slope of grain.

At the time of fixing, the moisture content should not exceed 24%. If the joists are preserved, it is normal to preserve the blocking etc. between the joists for the same risk category and to the same level as required to satisfy Table 4 of BS5268: Part 5: 1989 (see Table A9.1).

Battens fixed to a wall to support exterior cladding or interior panelling

Battens fixed on the outside surface of a wall to support exterior cladding of timber boards, and battens fixed to the interior surface of a wall to support interior panelling are normally of fairly small sections resawn from basic sawn sizes, with tolerances in accordance with BS4471: 1987. The most common species are European Whitewood or European Redwood, although British-grown or Canadian timbers can be used.

The timber should be free from decay and live insect attack, and any significant deformation which would require sections to be forced into place or lead to the vertical or horizontal 'line' of the battens not being sufficiently straight. The size of knots, splits and slope of grain should be limited. For convenience, rather than write an individual detailed specification, the specifier could decide that the battens should comply with the quality requirements for tiling battens in BS5534: Part 1: 1978 (1985) *Slating and tiling. Design*, including AMD3554 and AMD5781 (see Chapter A14).

A limit should be placed on the moisture content at the time of fixing. A suitable upper limit for battens on the exterior surface of a wall, extracted from design code BS5268: Part 2: 1988, would be 24%. The moisture content of battens to support interior cladding should be set by the guidance on service moisture contents given in Table A4.1.

As can be seen from Chapter A15, UK Building Regulations are not precise in stating whether or not exterior timber cladding should be preserved, therefore a similar uncertainty exists with regard to preserving the battens to support exterior cladding. The NHBC call for softwood battens to support exterior cladding to be preserved but the level of preservation is not specified by NHBC. If the decision is taken to preserve, the specifier has the choice of at least two documents to refer to in choosing a suitable specification. These are Table 5 ('external woodwork in buildings above the damp-proof course') of BS5589: 1989 on 'preservation of timber', or Table 4 ('timber in buildings') of BS5268: Part 5: 1989 in its reference to battens to support tiling battens. In the absence of specific guidance, the author is inclined to suggest the use of Table 4 of BS5268: Part 5 as guidance, particularly because this permits the use of CCA Schedules P8 and P9 as relevant for thin sections. If this is agreed, see Table A14.2. If, however, the decision is taken to use Table 5 of BS5589: 1989 as guidance, see Table A15.1. It should be remembered that, even in an area where House Longhorn Beetles are active, there is no requirement to preserve against them in a vertical wall, only in a 'roof void'.

A CCA or organic solvent preservative can be used.

Concealed non-structural timber framing

One encounters many cases where timber is used as a framing to support boards or boarding and is completely concealed once the work is complete. In such cases, although the appearance of the timber is not important, it should have enough strength, it should be sufficiently straight, it should be sufficiently free from splits, checks and shakes to allow retention of fixings, and it should be free from decay and live insect attack, and perhaps also free from wane. Although it is not the intention that BS1186: Part 3 on wood trim should apply to concealed timber framing, rather than produce an individual detailed specification the specifier could consider that the quality of framing should comply with the lowest Class of BS1186: Part 3 (i.e. Class 3), when this standard is published, but must make it clear which surfaces are to match 'Exposed' and 'Concealed' requirements of Part 3 (see Chapter B4).

Concealed framing is often made from small sections which are resawn from basic sawn sizes. The tolerances of BS4471: 1987 may be taken as a guide, although slightly larger deviations should not normally lead to problems.

Guidance on the moisture content requirement at the time of fixing can be obtained from Table A4.1. It is not normal for concealed framing for use in a dry environment to be preserved, other than perhaps if in a 'roof void' in an area where House Longhorn Beetles are active. If, however, the specifier decides to call for preservation, a decision will also be required on the most relevant treatment and schedules (see BS5589: 1989 or BS5268: Part 5: 1989, and Chapter B5).

A6 TIMBER FOR DOMESTIC T & G SOFTWOOD FLOORING

INTRODUCTION

The sizes, tolerances, species, face quality, moisture content and grading of the type of tongued and grooved softwood boarding used in domestic property and for similar loading situations are covered in detail in BS1297: 1987. The thicknesses to use to support domestic loading are covered in Approved Document A1/2 to the Building Regulations 1985 for England and Wales, and are likely to be repeated in BS8103: Part 3 when it is published (anticipated in 1990). BS8103: Part 3 is intended to be the timber section of 'The structural design of low rise buildings. Code of practice for timber floors and roofs in housing'. Details of fixing softwood flooring are included in BS8201: 1987 which is the code of practice for flooring of timber, timber products and wood based panel products.

A specification for normal tongued and grooved softwood flooring should include details of the following Items (a) to (d) which may vary in the form of the wording used depending whether it is written by a specifier or a developer.

(a) The species which is to be used

The species which is to be used or reference to the species permitted by BS1297: 1987 as listed in Table A6.1 which, in the correct thickness, satisfy the spans in Approved Document A1/2. There is no intention that tongued and grooved softwood flooring should be stress graded. The background, however, to the choice of species is based on the assumption that the grade described in BS1297: 1987 has a strength which approximates to the visually graded Special Structural (SS) grade of BS4978: 1988, and that any species which, at the SS level, has the same bending strength and stiffness (from BS5268: Part 2: 1988) as SS European Whitewood should be permitted to satisfy BS1297: 1987.

Table A6.1 Softwood species for flooring to satisfy BS1297: 1987

Imported European Redwood
Imported European Whitewood
British-grown Corsican Pine
British-grown Douglas Fir
British-grown Larch
British-grown Scots Pine
Canadian Douglas Fir-Larch
Canadian Hem-Fir
Canadian Spruce-Pine-Fir
USA Douglas Fir-Larch
USA Hem-Fir
USA Southern Pine

Note that, for conformity with BS1297: 1987, some species are included in Table A6.1 which are not otherwise dealt with in this book.

(b) The grade which is to be used

BS1297: 1987 describes a grade in detail and, for normal tongued and grooved softwood boarding, it should be adequate to refer to BS1297: 1987 when specifying grade. What is not stated in BS1297: 1987 nor in BS8201: 1987, but which may be included in BS8103: Part 3 when published, is that end splits should be cut off before boards are fixed. If the specifier wishes, such a clause could be added to a specification, usually as a workmanship clause. (The intention is to add such a clause in BS8201.)

If thicker boards are required, or if boards are required to support loads in excess of domestic loading, a design is required. In such a design it would normally be considered to be adequate to assume that boards graded as BS1297: 1987 have at least the same unit strength and stiffness as the visually graded SS grade of BS4978: 1988. (See BS5268: Part 2: 1988 for stress values.)

(c) The moisture content at the time of delivery and fixing

To satisfy BS1297: 1987, the 'normal' moisture content at the time of

Table A6.2 Average moisture content of interior joinery from BS1186: Part 1: 1986

For buildings with intermittent heating	15 ± 2%
For buildings with continuous heating providing room temperatures of 12–19°C	12 ± 2%
For buildings with continuous heating providing room temperatures of 20–24°C	10 ± 2%

Table A6.3 Average moisture content of floor boarding in service as given in BS8201: 1987

Unheated building	15% to 19%
Intermittent heating with a substantial drop in temperature between periods of heating	10% to 14%
Continuous heating with the temperature maintained day and night throughout the year at a reasonably constant level	9% to 11%
Underfloor heating	6% to 8%

delivery should not exceed 19%. The point is made in BS1297: 1987 that if the temperature in a room will be continuous at 12–19°C or higher, boards with a lower moisture content (at time of delivery and fixing) should be used. Although floor boards are not interior 'joinery' as defined in BS1186: Part 1: 1986 on the quality of timber in joinery, Table A6.2 which is based on BS1186: Part 1: 1986 gives guidance on the level of moisture content which can be anticipated in service. At the time of delivery and fixing, the moisture content of boards should be close to the service moisture content. Guidance on moisture content is also given in BS8201: 1987. The relevant parts are reproduced in Table A6.3. They do not tie up exactly with the figures from BS1186: Part 1: 1986.

BS8201: 1987 calls for softwood boards complying with BS1297 to be pulled together tightly with care during fixing. This implies that the moisture content at the time of fixing is close to and not significantly *lower* than the service moisture content. In the relatively rare case as shown in Fig. A6.1, however, where boards are fixed parallel to joists on noggings which are at right angles to the joists, extra care must be taken with checking moisture content and the extent to which boards are pulled together during fixing. This is to ensure that boards do not take up moisture after fixing, to the extent of leading to problems with expansion causing bursting apart at the ends of noggings, or to 'hogging' of the flooring.

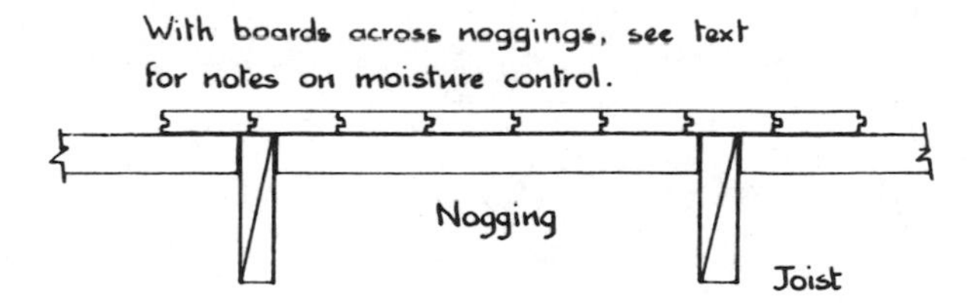

Fig. A6.1 Floor boards parallel to joists (see text)

Table A6.4 Dimensions of softwood tongued and grooved boards from BS1297: 1987[1]

Finished thickness (mm)[2]	A	16	19	21	28
Finished tongue thickness (mm)	B	4.5	6	6	6
Finished tongue top width (mm)	C	7	7	7	7
	D	7	7	8	12
Finished width of face[2]		65	90	113	137

[1] See Fig. A6.2 for positions of A, B, C and D.
[2] Note that any combination of width and thickness is acceptable in BS1297: 1987. Fixing requirements for various widths are detailed in BS8201: 1987, particularly Section Two.

(d) Sizes, profile and face quality

The sizes of softwood flooring boards as described in BS1297: 1987 are shown in Table A6.4 including the dimensions of the tongue. The profile type of the tongue is shown in Fig. A6.2. On the 'finished thickness' no minus tolerance is permitted by BS1297: 1987. A plus tolerance of 2 mm is permitted. On the face width the tolerance is ± 1 mm. Lengths of individual boards must be at least 1.8 m (although, presumably if they run across a width of less than 1.8 m or have to be cut to suit access points for services, they will have to be shorter).

Note that the finished sizes of softwood boards to comply with BS1297: 1987 are measured at the specified moisture content range and *not at 20%* as is the normal case for timber sizes complying with BS4471: 1987.

To comply with BS1297: 1987 there must be at least 10 cutter marks per 30 mm of length on an exposed face, although the underside may show 'hit-and-miss' planing.

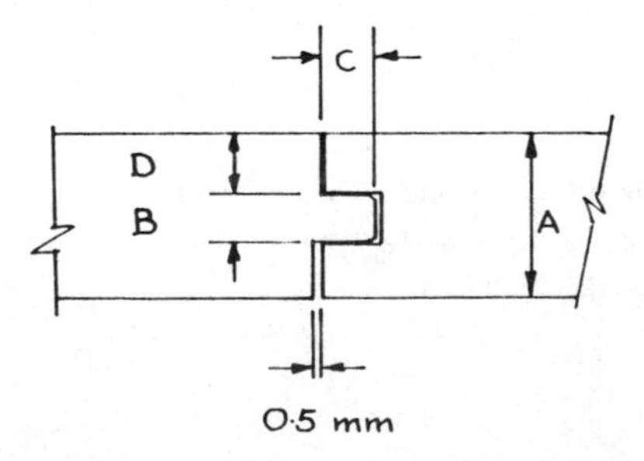

Fig. A6.2 Profile of tongued and grooved joint in BS1297: 1987

(e) Preservation

Although floor boards may be preserved, it is not normal for them to be preserved. (If tongued and grooved boards are used in a flat roof and need to be preserved, see Table 4 of BS5268: Part 5: 1989 on preservative treatment of structural timber, and Chapter A9 of this book for guidance.)

THICKNESSES REQUIRED

Approved Document A1/2 to the Building Regulations 1985 for England and Wales requires softwood tongued and grooved boards for domestic loading (not including the weight of partitions) in dwellings not exceeding three storeys to be at least:

16 mm thick if supported at a joist spacing of up to 450 mm and

19 mm thick if supported at a joist spacing of up to 600 mm.

TYPICAL SPECIFICATIONS

Each specifier of floor boarding must make the final decision on what and how to specify, but the following are given as two examples of the general form of typical specifications for softwood floor boards to support domestic floor loading in dwellings not exceeding three storeys. They do not cover all possible cases.

Typical specification for domestic floor boards of the type shown in BS1297: 1987 to span 600 mm between floor joists.
Tongued and grooved softwood floor boards are to be of European Whitewood and are required to have a finished thickness of at least 19 mm and a finished face width of 113 mm, with tongues and grooves, tolerances, grade and face quality of machining complying with BS1297: 1987. The average moisture content at time of fixing is to be between 10% and 14% with no reading higher than 19%. No length as delivered is to be less than 1.8 m. End splits are to be cut off before fixing. Fixing is to be as BS8201: 1987.

Typical specification for domestic floor boards of the type detailed in BS1297: 1987 to span 450 mm between floor joists.
Tongued and grooved softwood floor boards are required to have a finished thickness of at least 16 mm and a finished face width as defined in BS1297: 1987 of 90 mm. The tongues and grooves, tolerances, grade and face quality of machining are to be in accordance with BS1297: 1987. The species may be any species acceptable to BS1297: 1987. The average moisture content at time of fixing is to be between 10% and 14% with no reading higher than 19%. No length as delivered is to be less than 1.8 m. End splits are to be cut off before fixing. Fixing is to be in accordance with BS8201: 1987.

A7 TIMBER FOR TRUSSED RAFTER ROOFS

INTRODUCTION

The definition of trussed rafters is given in BS5268: Part 3: 1985 which is the code of practice for trussed rafter roofs, as 'light-weight triangulated frameworks spaced at intervals generally not exceeding 600 mm and made from timber members fastened together in one plane'.

A specification for the timber in a trussed rafter roof should include details covering Items (a) to (e) outlined below, which may vary in the form of wording used depending whether it is written by a specifier or a manufacturer. The specification should include details of the timber used in the trussed rafters, bracing, gable-ladders, wallplates, framing around openings wider than the centres between trussed rafters, timber in hip ends or changes in direction, and under water tanks. BS5268: Part 3: 1985 is very detailed and should normally be referenced in any specification of timber for trussed rafter roofs. When Part 3 of BS8103 for small domestic dwellings is published, it may also include details useful for a specification of the non-structural or semi-structural timbers in trussed rafter roofs (e.g. the quality of timber for wallplates), which are currently not included in BS5268: Part 3: 1985.

(a) The species which is to be used or a list of the species which are acceptable to the specifier.

Trussed rafters

By far the most common species used for trussed rafters is European Whitewood. A certain amount of European Redwood, Canadian Spruce-Pine-Fir, or British-grown Sitka Spruce is used. Canadian Hem-Fir is also used to a certain extent in the construction of trussed rafters but one would not normally specify Canadian Douglas Fir-Larch. If a species mix is used, the specifier must remember to exclude the use of structural finger joints (see Chapter B10).

Even if the required strength of a member is specified as a Strength Class (see Chapter B3) of design code BS5268: Part 2: 1988, it is advisable to specify the actual species which may be used, particularly if the timber is to be preserved or finger jointed.

Bracing

The minimum size of standard bracing members for trussed rafters spaced at centres not exceeding 600 mm is given in BS5268: Part 3: 1985 as 22 × 97 mm which has an influence on the species which is used. It is normal to use 22 × 100 mm European Whitewood but other species can be used. For example, if the trussed rafters are fabricated from Canadian Spruce-Pine-Fir, one could use this species mix for the bracing members providing that the timber is as wide (i.e. 97 mm) as required for the standard bracing, or if a special design is prepared to prove that narrower bracing is adequate.

Gable-ladders

If gable-ladders are pre-made by a manufacturer of trussed rafters or timber stud walls, it is quite normal to use the same species as in the trussed rafters or the stud walls. If gable-ladders are assembled on site, it is normal to use European Whitewood, European Redwood or Canadian Spruce-Pine-Fir.

Wallplates

If a 'wallplate' is part of a timber stud wall, the species is normally the same as for the stud wall (see Chapter A10). If it is placed on top of a masonry wall, the tendency is to use European Whitewood or European Redwood although other species of a suitable width and thickness, see Item (d) below, can be used.

Trimming around openings

When trimming around an opening wider than the distance between trussed rafters, or constructing a hip end or an infill at a change in direction, the species of timber used is normally the same as that in the trussed rafters.

Support for water tanks

If water tanks are supported on short timber joists, guidance on the size and stress grade is given in BS5268: Part 3: 1985, based on the use of European Whitewood or European Redwood.

(b) The stress grade, or grade which is to be used

As an alternative to a stress grade, a Strength Class may be specified (see Chapter B3).

All timber in trussed rafters, and structural timber in trussed rafter roofs must be stress graded. If the specification is being written by a specifier rather than a manufacturer, it is sensible to discuss with the manufacturer, or leave the manufacturer to select the most advantageous stress grade/species combination.

When European Whitewood or European Redwood is used, M75 (which matches Strength Class SC5 with European Whitewood/Redwood) is the most popular stress grade, but visually stress graded SS (which matches SC4 with European Whitewood/Redwood) is used as is MSS or M50. GS/MGS can be used but is not normally used.

If British-grown Sitka Spruce is used, it is normally stress graded M75 (which, for Sitka Spruce, matches Strength Class SC3, although BS5268: Part 3: 1985 permits span tables for M75 British-grown Sitka Spruce to have advantages over other timbers which match SC3).

The 1988 version of BS5268: Part 2 introduced additional stress grade/species combinations and tables of Canadian timber machine stress graded, some of which may be used for timber for trussed rafters (usually 38 mm thick). It is not yet clear which of these machine stress grades will be available in the UK, but one imagines that they will be at the level of Strength Classes SC4 or SC5.

The stress grade used for the framing around openings, for framing hip ends and at changes in direction will be determined by the designer in individual cases.

If water tanks are supported on short timber joists, guidance on what size to use is given in BS5268: Part 3: 1985 based on European Whitewood/Redwood stress graded GS/MGS.

Bracing, gable-ladders, wallplates

There is no requirement to stress grade the standard bracing specified in BS5268: Part 3: 1985 or to stress grade the timber in gable-ladders or wallplates although, if a wallplate is part of a timber stud wall, for convenience it is normal to use the same stress grade as for the studs (see Chapter A10). For standard bracing, gable-ladders, and wallplates on top of a masonry wall, normally it is adequate to call for the timber to be free from wane, decay and live insect attack, any significant distortion which would require it to be forced into place, and to be free from any major strength reducing feature. If a more detailed specification is required, see Chapter A5 for suggestions.

(c) The moisture content at time of fabrication and erection

If trussed rafters are fabricated with glued plywood gussets, the moisture content of the timber (and the plywood) at the time of fabrication should be as required by the adhesive manufacturer, but the moisture content of the timber must be not more than 20% (see BS6446: 1984 on the manufacture of glued structural components of timber and wood based panel products). With this exception, BS5268: Part 3: 1985 permits the moisture content at the time of fabrication of trussed rafters to be not more than 22%, which can also be taken as the upper limit for all timber in a trussed rafter roof. At the time of erection, the moisture content should not exceed 24% (see Chapter B2).

(d) The size and tolerances of members, and processing required

Trussed rafter members

The sizes of trussed rafter members are detailed in span tables in BS5268: Part 3: 1985 or in contract drawings, with tolerances on size also covered in clauses of BS5268: Part 3: 1985. It is normal for specifications to refer to BS5268: Part 3: 1985 and to the contract drawings. All trussed rafter material to satisfy BS5268: Part 3: 1985 is processed to close tolerances. In the case of European timbers it is normal to process both dimensions (but not to the same extent as planing of joinery timber). In the case of Canadian timber it is normal to use 'surfaced' CLS sizes (see Table B1.2). In either case, BS5268: Part 3: 1985 does not permit any minus tolerance on the specified finished size measured at 20% moisture content. Normally trussed rafter material for spans of up to 11 m is a dry finished thickness of 35 mm, particularly if on a domestic roof, although if Canadian CLS material is used, it is normally surfaced to a thickness of 38 mm. Designs for longer spans are based on thicker material.

BS8212: 1988 is the code of practice for 'dry lining and partitioning using gypsum plasterboard'. One general requirement is that the thickness of timber members (either with or without a batten on the side) to support a plasterboard ceiling, should be at least 41 mm where a butt joint occurs at the ends (but not the edges) of plasterboard sheets. One of the waivers included in BS8212: 1988 to this general requirement is for ceiling joists in trussed rafters designed and manufactured in accordance with BS5268: Part 3: 1985, where experience with many millions of trussed rafters has shown thicknesses of 35 mm to be adequate without the addition of a batten where a butt joint occurs at the ends of plasterboard sheets.

The size and tolerances of timber members to frame around openings is normally the same as the corresponding timber in the trussed rafters.

The size and tolerance of timber in hip ends and at changes of direction must be specified in individual cases by the designer.

Bracing

It is normal to use 22 × 100 mm basic sawn sizes for the standard bracing detailed in BS5268: Part 3: 1985. The minimum size permitted by BS5268: Part 3: 1985 for standard bracing is 22 × 97 mm. The tolerances would normally be expected to comply with BS4471: 1987 rather than BS5268: Part 3: 1985.

Gable-ladders

Gable-ladders can be made from sawn timber, either assembled on site by the builder or as a pre-made item. As an alternative to sawn timber, some manufacturers of trussed rafters offer a service of supplying pre-made gable ladders, in which case they may use basically either the same material as is used in the trussed rafters or, particularly if they also manufacture timber stud walls, basically the same material as is used in stud walls. In any of these cases, however, there is no requirement for the timber to be stress graded.

Support for water tanks

If water tanks are supported on timber members, it is normal to use basic sawn sizes of 47 mm thickness 'regularized' in depth (see the table of sizes in BS5268: Part 3: 1985). Typical regularized sizes are 47 × 72 mm, 47 × 97 mm, 47 × 120 mm or 47 × 145 mm. The tolerances of BS4471: 1987 apply.

Wallplates

If a 'wallplate' is the upper horizontal member of a timber stud wall, it is likely to be the same size as the members in the stud wall. If a timber wallplate is fixed to the top of a masonry wall, it is likely to be a piece of sawn timber at least 75 mm wide or 0.008 × span of the trussed rafter (which is guidance from BS5268: Part 3: 1985), whichever is the wider, normally at least 47 mm thick but possibly as thin as 38 mm. 75 × 100 mm is another size used on a masonry wall. The tolerances of BS4471: 1987 apply.

(e) The type of preservation, if required, and method of application

As can be seen from reference to Table B5.4, and considering the risk of insect attack in a normal 'dry' and ventilated trussed rafter roof as being 'negligible' or a 'low risk' (i.e. Risk category 1 or 2) and the risk of fungal attack as being even less, except in areas where House Longhorn Beetles are active (when all softwood in a 'roof void' must be preserved), or where there is a particular risk due to a local condition in a building, preservation of timber in a trussed rafter roof is normally 'unnecessary' or 'optional'. If, however, the specifier decides to opt for preservation, the strong inference in BS5268: Part 3: 1985 is to favour the use of an organic solvent rather than a water borne preservative. There are several reasons for this preference:

- A water borne preservative applied after processing of the timber (but before fabrication of the trussed rafters) is likely to affect the surface tolerances to a degree which may be unacceptable for the subsequent application of metal plate gussets (or glued plywood gussets).
- A water borne preservative increases the moisture content to such an extent that either kiln drying or air drying is required after preservation. There is rarely time to wait for air drying, and rarely a sufficient allowance in the price to cover kiln drying after preservation.
- An organic solvent is compatible with metal plate fasteners.

If an organic solvent is used to preserve against House Longhorn Beetles or common Furniture beetles, an insecticide must be included in the formulation.

If the decision is taken to preserve, normally it will not be necessary for a specifier to go to the extent of detailing the actual preservation schedules which can be used. On occasions, however, it may be necessary for a specifier to know if, for example, immersion or a double vacuum process is required, or if a particular species can be preserved with the equipment available, therefore guidance on schedules is given in Table A7.1. If an organic solvent is used, Table 4 of BS5268: Part 5:

Table A7.1 Organic solvent schedules and immersion periods for timber in trussed rafter roofs to satisfy BS5268: Part 5: 1989

Trussed rafters in a 'dry' pitched roof[5]	'Permeable' or 'Moderately resistant' species 'P' or 'MR' [3]	'Resistant' or 'Extremely resistant' species 'R' or 'ER'
In an area where House Longhorn Beetles are active (i.e. Risk 4 'Essential' for insect attack[1]) but where there is 'negligible risk' of fungal attack[1]	Schedule V/1 or 10 minute immersion[2]	Schedule V/1 or 10 minute immersion[2]
In a situation where common Furniture beetles are a 'low risk' (i.e. Risk 2 'Optional' for insect attack[1]) but it is decided to preserve against insect attack	Schedule V/1 or 3 minute immersion[2]	Schedule V/1 or 10 minute immersion[2]
In a situation where there is a 'low risk' of fungal attack[1] but it is decided to preserve against fungal attack	Schedule V/1 or 3 minute immersion	Schedule V/1 or 10 minute immersion[4]

See the text for cases where there is considered to be a risk of both fungal and insect attack.

[1] See Tables B5.4 and B5.5, and Tables 1, 2 and 3 of BS5268: Part 5: 1989 for further explanation and application of the 'Risk categories'.

[2] The organic solvent must include an insecticide if the risk is insect attack.

[3] When treating mixed species, see the relevant paragraph of Chapter B5 and the note below the second typical specification at the end of this chapter.

[4] When treating 'Resistant' or 'Extremely resistant' species against fungal attack, immersion is permitted for pitched roofs 'dry'.

[5] For preservation of wallplates see Table A5.1. For preservation of bracing use the schedules or immersion periods in this table. For preservation of tiling battens see Table A14.2.

1989 on the preservative treatment of structural timber, requires the schedules noted in Table A7.1 to be used in 'dry' conditions. Also note the schedules for cases where the specifier decides that there is a risk of fungal attack. If there is considered to be a risk both of fungal and insect attack, the more onerous of the treatments given in Table A7.1 should be used.

Trussed rafters are not normally suited for use in roofs where there is a risk of the timbers becoming 'wet', therefore no schedules for such conditions are given in Table A7.1.

Although not part of the specification for timber or preservation, it is worth noting that BS5268: Part 3: 1985 requires all nails in a trussed rafter roof to be hot-dip galvanized, Sherardized, or to be given another suitable treatment against corrosion.

TYPICAL SPECIFICATIONS

Each manufacturer or specifier of timber for trussed rafter roofs must make the final decision on what and how to specify, but the following are given as two examples of the general form of typical specifications. They do not cover all possible cases.

Typical specification for European timber in a trussed rafter roof, including bracing, timber in the gable-ladders and the wallplates. This roof is in an area where House Longhorn Beetles are a risk but there is not considered to be any other risk of insect or fungal attack. The trussed rafters are to be assembled using punched metal plate fasteners.
Timber in trussed rafters to be European Whitewood machine stress graded to M75 or to Strength Class SC5, in the processed sizes shown on Drawing _____ with tolerances on sizes complying with BS5268: Part 3: 1985. Timber in bracing to be basic sawn 22 × 100 mm (or 22 × 97 mm minimum if resawn from a wider piece) European Whitewood, free from significant distortion and free from any major strength reducing defect or wane. Tolerances on size are to be in accordance with BS4471: 1987. Timber in gable-ladders to be European Whitewood in the processed sizes as shown on Drawing _____ free from significant distortion and wane. Tolerances on size are to be in accordance with BS4471: 1987. Timber in wallplates to be basic sawn 75 × 100 mm European Whitewood or European Redwood with tolerances to BS4471: 1987, free from wane and significant distortion which would make it necessary to force the wallplates into place. At the time of fabrication, the moisture content of the timber is not to exceed 22%. At the time of erection, the moisture content of the timber is not to exceed 24%. All timbers are to be treated to comply with BS5268: Part 5: 1989 against House Longhorn Beetles with an organic solvent preservative containing an effective insecticide.

Typical specification for Canadian Spruce-Pine-Fir used in the fabrication of trussed rafters in a situation where preservation is considered to be unnecessary.
Timber in trussed rafters to be Canadian Spruce-Pine-Fir in the surfaced CLS sizes shown in Drawing ____ with tolerances complying with BS5268: Part 3: 1985. The stress grade/species combination is to satisfy Strength Class SC4 of BS5268: Part 2: 1988. Note that structural finger jointing (of mixed species) is not permitted. At the time of fabrication, the moisture content is not to exceed 22%. At the time of erection, the moisture content is not to exceed 24%.

In the second example above, it is assumed that it is 'unnecessary' to preserve the timber but, if a mixed species is preserved, it is normal to use a schedule designed for the most 'resistant' species. In the case of preserving Canadian Spruce-Pine-Fir with an organic solvent, however, this can lead to over-absorption, and it may be advantageous to agree the type of modified schedule normally used for preserving Spruce-Pine-Fir studs in the external walls of timber framed housing. (See Chapters A10 and B5, and NHBC Practice Note 5 or BS5268: Part 5: 1989.)

A8 TIMBER FOR TRADITIONAL 'CUT' PITCHED ROOFS

INTRODUCTION

Trussed rafter roofs are so well established in the UK that they could be regarded as being 'traditional'. However, this chapter is intended to deal with the type of roof shown in outline cross-section in Fig. A8.1, which is usually referred to as a traditional 'cut' roof. Such roofs are constructed from the timber components, design sizes of which used to be covered in 'Schedule 6' of previous Building Regulations and are now covered by span tables in Approved Document A1/2 to the Building Regulations 1985 for England and Wales. It is likely that similar tables will be produced and will link with the Scottish Building Standards. It is also the intention to produce similar tables to be presented in Part 3 of the code for small dwellings (BS8103: Part 3) when printed.

The tables cover, or will cover, basic sawn softwood sizes or regularized sizes (see Fig. B1.2) which comply with the tolerances of BS4471: 1987, and 38 mm thick 'surfaced' CLS/ALS sizes (see

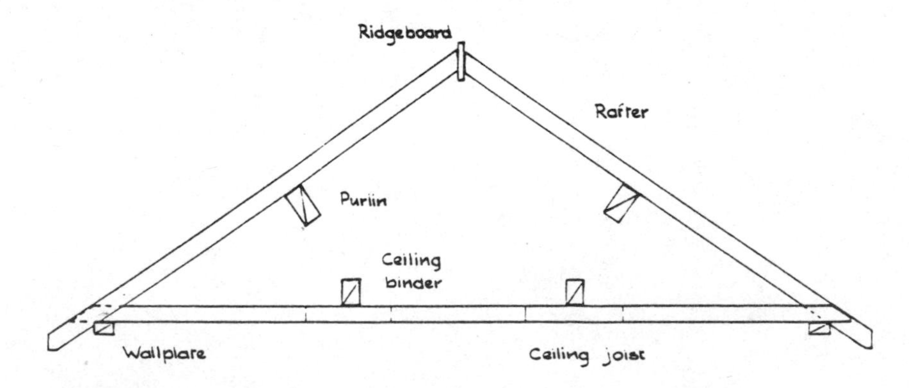

Fig. A8.1 Cross-section through a typical traditional 'cut' pitched roof

Chapter B1) which also comply with tolerances in BS4471: 1987. The intention is that the span tables in BS8103: Part 3, and future tables linked to the Building Regulations for England and Wales will quote 'regularized' sizes rather than sawn sizes (and will quote CLS/ALS sizes) where a ceiling is to be fixed directly to the underside of a member.

Although intended primarily to specify requirements to be met in the production of load/span tables, a method of designing timbers in traditional 'cut' roofs is detailed in several Sections of BS5268: Part 7 which are either published or in the process of being published by BSI. The numbers of the four Sections which are relevant to 'cut' roofs are listed below in Item (d).

A specification for the timber in a traditional cut roof should include details covering Items (a) to (e) outlined below, which may vary in the form of the wording used depending whether it is written by a specifier or a builder. The specification should include details of the timber used for the common rafters, jack rafters (if applicable), purlins supporting rafters, ceiling joists, binders supporting ceiling joists, ridge boards and wallplates.

(a) The species which is to be used or a list of the species which are acceptable to the specifier

(b) The stress grade which is to be used

As an alternative to quoting a species and a stress grade, a Strength Class (see Chapter B3) can be specified for the structural timber. Each Strength Class in design code BS5268: Part 2: 1988 can be satisfied by several stress grade/species combinations as can be seen in BS5268: Part 2. A small selection is shown in Table B3.3.

If the specification for structural timber is based on a Strength Class, care must be taken to ensure that the various stress grade/species combinations which could be supplied satisfy all the functional criteria. As well as checking that the possible combinations satisfy the strength requirements, it is necessary to check or specify that the timber will be available sufficiently dry, that the sizes and tolerances will satisfy the design size, that if required to be preserved, the timber can be preserved to the required level, and that there are no particular restrictions on the use of fasteners or connections.

The various span tables for rafters etc. in Approved Document A1/2 which support the Building Regulations 1985 for England and Wales are based on Strength Classes SC3 and SC4 of BS5268: Part 2: 1988.

Species commonly used for rafters and ceiling joists are European Whitewood, European Redwood or Canadian Spruce-Pine-Fir, with British-grown Sitka Spruce also used. Canadian Spruce-Pine-Fir to satisfy the span tables linked to Building Regulations must be either 'surfaced' CLS 38 mm thick, or be sawn or resawn to the BS4471: 1987 sizes and tolerances. Timber sawn to NLGA sizes and tolerances can be used, but must be designed as the correct dry design size, not as the green sawn size, nor assumed to be one of the sawn sizes in the tables in Approved Document A1/2. Similar comments on special sizes apply for purlins supporting rafters and binders supporting ceiling joists, except that they are normally thicker than 38 mm.

Ridgeboards are normally European Whitewood or European Redwood, but British-grown Sitka Spruce or Canadian Spruce-Pine-Fir or Hem-Fir can be used. No British Standard covers the grade of timber required for timbers such as ridgeboards. It may be sufficient to quote a sawmill grade such as Nordic 'fifth' quality, but such a form of specification has the disadvantages as explained in Chapter B4 and A5. Normally it is adequate to call for the timber to be free from wane, decay and live insect attack, and to be free from any significant distortion which would require it to be forced into place. If a more detailed specification is required, see Chapter A5 for suggestions.

If a wallplate is part of a timber stud wall, the species and stress grade will normally be the same as the timber in the stud wall (see Chapter A10), even though there is no specific requirement for the wallplate to be stress graded. If a wallplate is placed on top of a masonry wall, the tendency will be to use European Whitewood or European Redwood although other species of suitable width and thickness (see Item (d) below) can be used. For a wallplate on top of a masonry wall, normally it is adequate to call for the timber to be free from wane, decay and live insect attack and to be free from any significant distortion which would require it to be forced into place. If a more detailed specification is required, see Chapter A5 for suggestions.

(c) The moisture content at the time of erection

To comply with design code BS5268: Part 2: 1988, the moisture content at the time of erection should not exceed 24% (see notes below Table B2.2).

(d) The size and tolerances of members, and processing required

If the stress graded timbers in a traditional cut roof (i.e. common rafters, jack rafters, purlins supporting rafters, ceiling joists and binders supporting ceiling joists) are to comply with the span tables in Approved Document A1/2 to the Building Regulations 1985 for England and Wales, the sizes must be as tabulated in Approved

Document A1/2. Note that these are currently all for timber of the basic sawn sizes and tolerances of BS4471: 1987, or for North American CLS/ALS sizes 38 mm thick, not for 'regularized' sizes (see Chapter B1). The Regulations make it clear that regularized sizes must not be used in satisfying these particular tables.

It is the intention that future span tables of this type will be based on the guidance given in the various Sections of BS5268: Part 7 (see a list below of the relevant Section numbers), and this will require a complete recalculation of sizes. When that occurs, it is likely that the principle will be adopted of quoting regularized sizes (and CLS/ALS sizes) where a member is a direct support for a ceiling. Certainly that decision has been taken by the relevant BSI committees for the tables to be presented in the timber Part of the code for small buildings (BS8301: Part 3). These tables may be repeated in some form of document 'approved' to link with the Building Regulations for England and Wales, or BS8301: Part 3 may become 'approved'.

Despite the notes in the two previous paragraphs which refer to official published span tables for timbers in 'cut' roofs, any sawn, regularized, 'surfaced' or planed size can be used provided that an individual design is carried out in accordance with the design method given in the relevant Section of BS5268: Part 7. The Sections which apply to cut roofs are:

Section 7.3: 1989 *Ceiling joists*,
Section 7.4: 1989 *Ceiling binders*,
Section 7.5: 19__ *Rafters*,
Section 7.6: 19__ *Purlins supporting rafters.*

It is anticipated that Sections 7.5 and 7.6 will be published in 1990.

BS8212: 1988 is the code of practice for 'dry lining and partitioning using gypsum plasterboard'. One general requirement is that the thickness of timber members (either with or without a batten on the side) to support a plasterboard ceiling, should be at least 41 mm where a butt joint occurs at the ends (but not the edges) of plasterboard sheets. One of the waivers included in BS8212: 1988 to this general requirement is for joists in domestic situations, where experience over many years has shown thicknesses of 38mm to be adequate without the addition of a batten where a butt joint occurs at the end of plasterboard sheets.

If a wallplate is the upper horizontal member of a timber stud wall, it is likely to be surfaced to 38 × 89 mm. If a timber wallplate is fixed to the top of a masonry wall, it is likely to be a piece of sawn timber of at least 75 mm basic width and normally of at least 47 mm thickness although possibly as thin as 38 mm. 75 × 100 mm is another size used on top of a masonry wall.

A ridgeboard in a domestic size building is likely to be a sawn section with a basic thickness of 22 or 25 mm, and a basic depth of 150, 175, 200 mm or deeper depending on the size and slope of the incoming rafters.

(e) The type of preservation, if required, and method of application

As can be seen from reference to Table B5.4, and considering the risk of insect attack in a normal 'dry' and ventilated traditional cut pitched roof as being 'negligible' or a 'low risk' (i.e. Risk category 1 or 2) and the risk of fungal attack to be even less, except in areas where House Longhorn Beetles are active (when all softwood in the 'roof void' must be preserved), or where there is a particular risk due to a local condition in a building, preservation of timber in a traditional cut pitched roof is normally 'unnecessary' or 'optional'. If, however, the specifier decides to opt for preservation, although the choice may well be to use an organic solvent, a water borne CCA preservative can be used and is not regarded as having the same disadvantages as quoted for a trussed rafter roof, see Item (e) of Chapter A7.

If an organic solvent is used to preserve against House Longhorn Beetles or common Furniture beetle, an insecticide must be included in the formulation.

If the decision is taken to preserve, normally it will not be necessary for a specifier to go to the extent of detailing the actual preservation schedules which may be used. On occasions, however, the specifier may wish to state if a water borne or organic solvent preservative is to be used and, in the case of an organic solvent, to know if immersion is permissible. Guidance is given, therefore, in Table A8.1 of this chapter on schedules which satisfy Table 4 of BS5268: Part 5: 1989.

Although not part of the specification of timber or preservation, because a particular point is made in Chapter A7 of the requirement in BS5268: Part 3: 1985 that all nails in a trussed rafter roof should be treated against corrosion, it is worth noting the slight contradiction that there is no such requirement for a traditional cut pitched roof, even if a nail is performing the same function as in a trussed rafter roof.

BRACING

In the case of traditional cut pitched roofs of domestic size with tiling or slating battens fixed at close centres to the rafters, it is not normally considered to be necessary to provide the type of diagonal and longitudinal bracing which has been found to be necessary in a trussed rafter roof.

Table A8.1 Treatment schedules and immersion periods for traditional 'cut' pitched roofs to satisfy BS5268: Part 5: 1989

Timber in a 'dry' traditional 'cut' pitched roof[5]	'Permeable' or 'Moderately resistant' species 'P' or 'MR'[3]		'Resistant' or 'Extremely resistant' species 'R' or 'ER'	
	Organic solvent	CCA water borne	Organic solvent	CCA water borne
In an area where House Longhorn Beetles are active (i.e. Risk 4 'Essential' for insect attack[1]) where there is 'negligible risk' of fungal attack[1]	Schedule V/1 or 10 minute immersion[2]	Schedule P3 20 g/l solution	Schedule V/1 or 10 minute immersion[2]	Schedule P3 30 g/l solution
In a situation where common Furniture beetles are a 'low risk' (i.e. Risk 2 'Optional' for insect attack[1]) but it is decided to preserve against insect attack	Schedule V/1 or 3 minute immersion[2]	Schedule P2 20 g/l solution	Schedule V/1 or 10 minute immersion[2]	Schedule P2 30 g/l solution
In a situation where there is a 'low risk' of fungal attack[1] but it is decided to preserve against fungal attack	Schedule V/1 or 3 minute immersion	Schedule P2 20 g/l solution	Schedule V/1 or 10 minute immersion[4]	Schedule P2 30 g/l solution

See the text for cases where there is considered to be a risk of both fungal and insect attack.

1 See Tables B5.4 and B5.5 and Tables 1, 2 and 3 of BS5268: Part 5: 1989 for further explanation and application of the 'Risk categories'.

2 The organic solvent must include an insecticide if the risk is insect attack.

3 When treating mixed species, see the relevant paragraph of Chapter B5 and the note below the second typical specification at the end of Chapter A7.

4 When treating 'Resistant' or 'Extremely resistant' species by an organic solvent against fungal attack, immersion is permitted for pitched roofs 'dry'.

5 For preservation of wallplates see Table A5.1. For preservation of tiling battens see Table A14.2.

TYPICAL SPECIFICATIONS

Each specifier or builder of timber in a traditional cut pitched roof must make the final decision on what and how to specify, but the following are given as two examples of the general form of typical specifications. They do not cover all possible cases.

Typical specification for rafters, ceiling joists, purlins, binders, ridgeboards and wallplates in a traditional 'cut' pitched roof. The roof is not in an area where House Longhorn Beetles are active, but the assumption made is that the roof timbers are to be preserved against the possibility of fungal or insect attack.

Rafters, purlins supporting rafters, ceiling joists, binders supporting ceiling joists, ridgeboards and wallplates are to be the sizes shown on Drawing ____ with tolerances to BS4471: 1987. The stress grade/species combination used is to match Strength Class SC4 of BS5268: Part 2.

The timber is to be treated with an organic solvent or CCA water borne preservative against the possibility of fungal or insect attack to satisfy BS5268: Part 5: 1989 in an area where House Longhorn Beetles are not active.

At the time of delivery to site and erection, the moisture content of the timber is not to exceed 24% even if a water borne preservative is used.

Typical specification for European Whitewood timbers in a traditional cut pitched roof in an area where House Longhorn Beetles are active but where there is not considered to be a risk of fungal attack or other insect attack.

Rafters, purlins supporting rafters, ceiling joists, binders supporting ceiling joists, ridgeboards and wallplates to be European Whitewood.

The rafters, purlins, ceiling joists and binders are to be the sizes shown on Drawing ____ with tolerances in accordance with BS4471: 1987, stress graded to the SS or MSS rules of BS4978: 1988 (i.e. to satisfy Strength Class SC4 of BS5268: Part 2: 1988). Note that the rafters, purlins and binders are sawn sizes, the ceiling joists regularized sizes.

The ridgeboards are to be at least a basic sawn size of 22 × 150 mm, and the wallplates are to be a basic sawn size of 47 × 100 mm, both to be free from wane, decay, live insect attack, and free from any significant distortion which would require them to be forced into place.

The timber is to be treated with a CCA preservative complying with BS4072: Part 1: 1987 to the requirements of BS5268: Part 5: 1989 for an area where House Longhorn Beetles are active but where there is not considered to be a risk of other insect attack or of fungal attack.

At the time of delivery to site and at the time of erection, the moisture content is not to exceed 24%.

A9

TIMBER FOR FLOOR AND FLAT ROOF JOISTS

AND T & G SOFTWOOD BOARDING FOR FLAT ROOFS

INTRODUCTION

There is no British Standard which deals specifically with floor joists, flat roof joists or solid softwood tongued and grooved boarding for flat roofs. BS5268: Part 2: 1988 is the structural design code which must be satisfied, and reference can also be made to Sections 7.1 and 7.2 (see below) of BS5268. Also, it is the intention that BS8103: Part 3, when published as a code of practice to cover structural and semi-structural timber in small dwellings, will provide a specification and span tables for timber normally used for floors and flat roof joists in most dwellings. It is also the intention that BS8103: Part 3 will include span tables and a specification for solid softwood tongued and grooved boards for use on flat roofs, based on the use of the tongued and

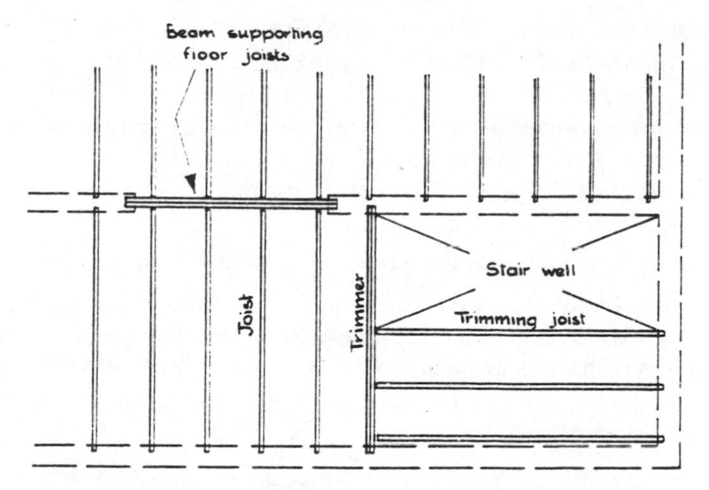

Fig. A9.1 Floor joists, trimmer joist, beam and trimming joist

grooved boards detailed in BS1297: 1987 for softwood flooring, and will provide span tables for a limited range of beams supporting floor joists, trimmer joists and trimming joists (see Fig. A9.1).

Currently the span tables for floor joists and flat roof joists in Approved Document A1/2 to the Building Regulations 1985 for England and Wales only cover timber of the basic sawn sizes and tolerances of BS4471: 1987, and 38 mm thick Canadian timber surfaced to CLS/ALS sizes with BS4471 tolerances. It does not cover 'regularized' sizes (see Chapter B1). It is the intention that, when BS8103: Part 3 and future versions of documents 'approved' to satisfy Building Regulations are published, the span tables will include regularized sizes and CLS/ALS sizes (but not basic sawn sizes) for the cases where a member is a direct support for a ceiling. Span tables in BS8103: Part 3 and any future versions of documents approved to satisfy Building Regulations will be calculated on the basis defined in:

BS5268: Part 7: Section 7.1: 1989 *Floor joists* and
BS5268: Part 7: Section 7.2: 1989 *Flat roof joists*.

When designing and specifying a joist, it is important to define any processing, and the tolerances permitted on size. The most readily available joists are:

(1) European timber sawn or regularized to the sizes and tolerances of BS4471: 1987 (i.e. virtually no minus tolerance permitted—see Chapter B1), stress graded to BS4978: 1988 on softwood grades for strucural use, to satisfy either the SC3 or SC4 Strength Class levels of design code BS5268: Part 2: 1988 (see Chapter B3). The majority of imported timber in this category is normally available dried to a moisture content of about 18/20/22%.

(2) Canadian timber surfaced to 38 mm thick CLS sizes with tolerances as in Appendix A of BS4471: 1987 (i.e. no minus tolerance permitted), stress graded to a NLGA (National Lumber Grading Association) grade which will satisfy either the SC3 or SC4 Strength Class levels of BS5268: Part 2: 1988. The majority of this timber (in Spruce-Pine-Fir) is normally available dried to a moisture content of 19% or less.

(3) Canadian timber sawn to NLGA Imperial rough sawn sizes with NLGA tolerances (often referred to as 'nominal' sizes within the UK timber trade—see Chapter C5), stress graded to a NLGA grade which will satisfy either the SC3 or SC4 Strength Class levels of BS5268: Part 2: 1988. This timber is normally imported 'green' (i.e. undried).

(4) More recently, an increasing percentage of Canadian 'green' sawn

timber is being sold so that, when dried to 20% moisture content, it will comply with the BS4471 sizes and tolerances. Some is stress graded to a NLGA grade, some to BS4978: 1988, to satisfy either the SC3 or SC4 levels of BS5268: Part 2: 1988.

The span tables in the current Approved Document A1/2 cover joist types as described in (1), (2), and (4) above at the SC3 and SC4 levels, but not Type (3). The joists described in (3) above can be used to satisfy BS5268: Parts 2 and 7 but, to do so, designs must be produced using the correct dry design sizes.

Some joist material is imported green. Some British-grown timber is sold green. BS5268: Part 2: 1988 requires joists and solid timber boarding for flat roofs to have a moisture content not exceeding 24% at the time of erection. It is preferable to dry joists to about 18/20% moisture content before they are loaded, or else excess deflection may well occur as the joists dry out under load.

Except for the timber in 'roof voids' in an area where House Longhorn Beetles are active, there is hardly a case in new domestic construction with good detailing where the preservation of joists in floors or roofs, or boarding on flat roofs can be said to be 'essential'. However, NHBC call for flat roof joists, and softwood boarding if used, to be preserved. NHBC also require at least the ends of floor joists to be preserved where the ends are built into solid (non-cavity) walls. Where joists are used as part of a suspended timber ground floor, the general recommendation of UK timber associations is to preserve the joists (but not the floor boarding) as an insurance against possible problems.

When considering whether or not to preserve, on the assumption that the building is well designed, it is normal to consider that the risk of fungal or insect attack (see Table B5.4) in a normal intermediate floor is 'negligible'; that there is only a 'low risk' of insect attack in a flat roof but that it is probably 'desirable' to preserve flat roof joists and solid boarding (if used) against fungal attack; and that it is 'desirable' to preserve ground floor joists against possible fungal and insect attack.

A specification of timber for floor joists, flat roof joists and solid boarding on flat roofs should include details covering Items (a) to (e) below, which may vary in the form of the wording used depending whether it is written by a specifier or a builder. The overall specification should include details of the wallplates, solid blocking or herringbone strutting between joists, and also of timber floor boarding (see Chapter A6), and timber boarding on flat roofs (if used). There is no specific British Standard for timber boarding on flat roofs. It is likely that, in the absence of a specific standard, BS8103: Part 3 (when published) will recommend that, if timber boarding is used on a flat roof, the same boarding should be used as is specified for a domestic floor in BS1297:

1987 (see Chapter A6), preserved if the flat roof joists are preserved. For the specification of wallplates, refer to the detailed notes in Chapters A5 and A8, and Table A5.1. For the specification of solid blocking or herringbone strutting between joists, refer to Chapter A5.

(a) The species which is to be used or a list of the species which are acceptable to the specifier, and

(b) The stress grade for joists, and the quality of boarding which is to be used

As an alternative to a stress grade/species combination for joists, a Strength Class can be specified (see Chapter B3). Each Strength Class in BS5268: Part 2: 1988 can be satisfied by several species/stress grade combinations as can be seen in BS5268: Part 2: 1988. A small selection is shown in Table B3.3.

If the specification of joists is based on a Strength Class, care must be taken to ensure that the various combinations which could be supplied satisfy all the functional criteria. As well as checking that the possible combinations satisfy the Strength Class, it is necessary to check that the timber will be available sufficiently dry, that the dry sizes and tolerances satisfy the design, and that, if required to be preserved, the timber can be preserved to the required level.

Species normally used for joists are European Whitewood or European Redwood, Canadian Spruce-Pine-Fir or Hem-Fir, or British-grown Sitka Spruce.

For the species and quality of solid tongued and grooved softwood boarding spanning between joists on flat roofs, see Chapter A6 for softwood flooring (and BS8103: Part 3 when published).

(c) The moisture content at the time of erection

If the specifier wishes to avoid the risk of unnecessary deflection taking place due to an initial period of drying-out under load, it is necessary to specify that the joists should be installed at a moisture content of about 18/20% or less. BS5268: Part 2: 1988 sets an upper limit at the time of erection of joists and flat roof boarding of 24% moisture content.

(d) The size, type of processing, and tolerances on size

Table B1.1 gives a list of basic sawn sizes as presented in BS4471: 1987, and Table B1.2 gives a list of surfaced CLS sizes as given in Appendix A of BS4471: 1987. When using sizes from the main body of BS4471:

1987 for joists supporting a ceiling, it is becoming more normal to regularize the depth. BS4471: 1987 calls for the regularizing reduction to be a maximum of 3 mm for depths (widths) up to 150 mm, and 5 mm for depths over 150 mm. These are the regularizing reductions which are normally used. Although BS4471: 1987 does not require the regularizing to be carried out when the timber is dry, it makes sense to do so, particularly because the 1987 version of BS4471 does not permit any minus tolerance on the regularized depth when measured at 20% moisture content. When specifying, it is necessary to clarify if a sawn or regularized size is required, e.g.

47 × 200 mm sawn,
47 × 195 mm regularized, or
47 × 195 mm regularized from 47 × 200 mm.

The BS4471: 1987 tolerances on sawn sizes not exceeding 100 mm are:

minus 1 mm, plus 3 mm and,

on sizes over 100 mm are:

minus 2 mm, plus 6 mm.

Minus tolerances are permitted on only 10% of pieces.

No minus tolerances are permitted on the processed dimension of regularized sizes, or CLS.

If green rough sawn NLGA sizes are specified, it is necessary to take note of the minus tolerances, and to make an allowance for the timber being dried to 20% moisture content in calculating the dry design size (see Table C5.3).

BS8212: 1988 is the code of practice for 'dry lining and partitioning using gypsum plasterboard'. One general requirement is that the thickness of timber members (either with or without a batten on the side) to support a plasterboard ceiling, should be at least 41 mm where a butt joint occurs at the ends (but not the edges) of plasterboard sheets. One of the waivers included in BS8212: 1988 to this general requirement is for joists in domestic situations, where experience over many years has shown thicknesses of 38 mm to be adequate without the addition of a batten where a butt joint occurs at the ends of plasterboard sheets.

(e) The type of preservation, if required, and method of application

As can be seen from reference to Table B5.4 and the notes above,

Table A9.1 Treatment schedules and immersion periods for floor joists, flat roof joists, solid timber decking etc. to satisfy BS5268: Part 5: 1989

	'Permeable' and 'Moderately resistant' species 'P' or 'MR' [3,7]		'Resistant' and 'Extremely resistant' species 'R' or 'ER'	
	Organic solvent	**CCA water borne**	**Organic solvent**	**CCA water borne**
Flat roof joists etc. in an area where House Longhorn Beetles are active (i.e. Risk 4 'Essential' for insect attack[1]) where there is 'negligible risk' of fungal attack[1]	V/1 or 10 minute immersion[2]	P3 20 g/l solution	V/1 or 10 minute immersion[2]	P3 30 g/l solution
Flat roof (cold) joists etc. in an area where House Longhorn Beetles are not active but it is decided to treat against the possible risk of fungal or insect attack[8]	V/2[2,6]	P3 20 g/l solution	V/4[2,6]	P7 30 g/l solution
Flat roof (warm) joists etc. in an area where House Longhorn Beetles are not active but it is decided to treat against the possible risk of fungal or insect attack[8]	V/1[2,6]	P3 20 g/l solution	V/4 or V/3[2,6]	P7 30 g/l solution
Ground floor joists etc.[4]	V/1 or 10 minute immersion[5]	P3 20 g/l solution	V/4 or V/3[2,6]	P7 30 g/l solution

[1] See Tables B5.4 and B5.5 and Tables 1, 2 and 3 of BS5268: Part 5 for further explanation and application of the 'Risk categories'.
[2] The organic solvent must include an insecticide if the risk is insect attack.
[3] When treating mixed species, see the relevant paragraph of Chapter B5.
[4] On the assumption that the sub-floor ventilation is adequate.
[5] Immersion permitted only for moderately resistant Pinus species (i.e. only for European Redwood of the species covered by this Book).
[6] According to BS5268: Part 5 (Table 4), immersion is not permitted for these situations.
[7] For preservation of wallplates see Table A5.1.
[8] Where there is likely to be high humidity, see Table 4 and notes of BS5268: Part 5: 1989.

except in areas where House Longhorn Beetles are active or where there is a particular risk due to a local condition in a building, preservation of joists is not essential. As stated above in the introduction, however, NHBC call for joists and boarding in flat roofs to be preserved, and timber associations in the UK suggest that joists (but not floor boarding) in timber suspended ground floors should be preserved.

If the decision is taken to preserve, normally it will not be necessary for a specifier to go to the extent of detailing the actual preservation schedules which may be used. On occasions, however, the specifier may wish to state if a water borne or organic solvent preservative is to be used and, in the case of an organic solvent, to know if immersion is permissible. Guidance is given, therefore, in Table A9.1 of this chapter on schedules which satisfy Table 4 of BS5268: Part 5: 1989.

TYPICAL SPECIFICATIONS

Each specifier of timber joists for floors or flat roofs, and tongued and grooved softwood boarding for flat roofs must make the final decision on what and how to specify, but the following are given as the general form of four typical specifications. They do not cover all possible cases.

Typical specification for intermediate floor joists, trimmer joists and trimming joists supporting domestic loading and with a plasterboard ceiling.
Floor joists, trimmer joists and trimming joists to be European Whitewood 47 × 195 mm regularized from 47 × 200 mm with tolerances in accordance with BS4471: 1987, stress graded to BS4978: 1988 to the SC4 level of BS5268: Part 2: 1988. At the time of regularizing, the moisture content is not to exceed 20%. At the time of erection, the moisture content is not to exceed 24%.

Typical specification for flat roof joists supporting a plasterboard ceiling in an area where House Longhorn Beetles are active.
Flat roof joists to be Canadian Spruce-Pine-Fir, 1⅞ inch × 8 inch green sawn timber with NLGA tolerances, regularized at a moisture content not exceeding 20% to a depth of 189 mm, stress graded to the NLGA or BS4978: 1988 rules to match the SC3 level of BS5268: Part 2: 1988. The joists are to be preserved by CCA or an organic solvent double vacuum process to satisfy BS5268: Part 5: 1989 against the possibility of fungal attack and House Longhorn Beetles. At the time of erection, the moisture content is not to exceed 24%.

Typical specification for ground floor joists supporting domestic loading.
Ground floor joists to be 47 × 225 mm basic sawn size with tolerances to BS4471: 1987. The stress grade/species combination is to match the SC4 level of BS5268: Part 2: 1988. Joists to be preserved to satisfy BS5268: Part 5: 1989 against the possible risk of fungal and insect attack.

Typical specification for tongued and grooved softwood boarding for use on joists on a flat roof.
Tongued and grooved softwood boarding on flat roof joists is to be European Whitewood of 19 mm finished thickness and with a finished width of face of 90 mm. The profile of boards, quality and tolerances is to be in accordance with BS1297: 1987. The boarding is to be preserved to satisfy BS5268: Part 5: 1989 against the possibility of fungal attack. At the time of fixing, the moisture content is not to exceed 24%.

A10

TIMBER FOR LOADBEARING STUD WALLS

INTRODUCTION

A specification for timber for loadbearing stud walls (see Fig. A10.1) should include details of the following items, which may vary in the form of the wording depending whether it is written by a specifier or a manufacturer. For non-loadbearing stud walls, either this chapter or the parts of Chapter A5 which refer to non-structural framing can be used for guidance.

(a) The species or species mix which is to be used, or a list of the species which are acceptable to the specifier

The majority of UK stud walls are manufactured from the Canadian species mix, Spruce-Pine-Fir (SPF). Some are manufactured from European Whitewood. Canadian Hem-Fir used to be used in large

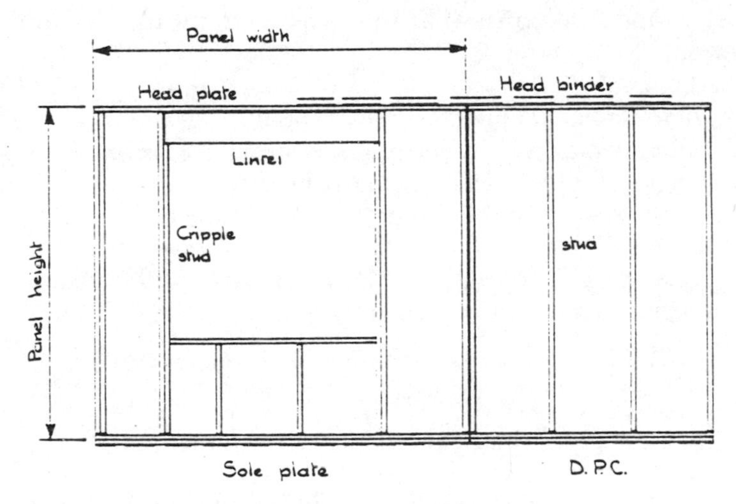

Fig. A10.1 Elevation on a typical exterior stud wall

quantities but it is normally only available undried ('green'), therefore there has been a move towards using dried SPF or European Whitewood.

(b) The stress grade to be used

Although a species and stress grade are often specified, as an alternative, it is possible to specify a Strength Class from the structural design code BS5268: Part 2: 1988 (see Chapter B3). If doing so, however, the specifier should ensure that the species permitted to be used are suited for the properties required in addition to strength (e.g. amenability to preservation, ease of nailing, dryness etc.).

The vertical studs in a loadbearing stud wall (see Fig. A10.1) should be stress graded. Although it is considered necessary to stress grade the 'cripple' studs, and it is normal practice also to stress grade the horizontal rails, it could be argued that it is not absolutely necessary to stress grade the rails because, to a large extent, they are only transmitting bearing. Noggings need not be stress graded. The members acting as lintols over openings should be stress graded.

If Canadian timber is used, the stress grade is likely to be a NLGA grade (see Chapter C5) although some importers of Canadian timber stress grade it to BS4978: 1988 (see Chapter B3). If European timber is used, unless the stud wall is assembled abroad, the timber is almost certain to be stress graded to BS4978: 1988. It is normal for loadbearing studs to be stress graded to a level not in excess of Strength Class SC4 of BS5268: Part 2: 1988, and a large percentage can be at the SC3 level.

(c) The moisture content at the time of manufacture and erection

At the time of manufacture, the timber should be at a moisture content close to that which will be encountered in service, i.e. an average of 14% m.c. (see Table B2.2). Particularly because of the speed of timber framed construction and the short period of time on site before stud walls are 'lined' on the inside surface, it is most unwise to use undried timber. At the time of erection, BS5268: Part 2: 1988 requires the timber to have a moisture content not in excess of 21% (see notes below Table B2.2).

(d) The size, type of surface, and tolerances on size

Because of the way in which current designs of timber framed constructions were introduced to the UK, the majority of stud walls on the exterior walls of dwellings are based on the use of 38 × 89 mm which is a Canadian Lumber Standards (CLS) size (see Chapters B1 and C5). Similarly, most interior stud walls in the UK are based on the use of 38 × 63 mm or 38 × 89 mm CLS. Because most UK drawings and designs are based on these sizes, some Nordic sawmills produce a version of CLS in these and larger sizes (but stress graded to BS4978: 1988).

It is normal for all surfaces to be 'surfaced' (see Chapter B1). To comply with Appendix A of BS4471: 1987 on sawn and processed softwood, no minus tolerance is permitted on processed dimensions of CLS at 19% moisture content. Plus tolerances are not limited but are normally small to ease the assembly into construction jigs at manufacturing units. Normally corners are rounded, but not more than 3 mm.

As an alternative to the use of 38 × 89 mm and 38 × 63 mm CLS, some UK manufacturers use sizes processed from sawn sizes selected from the main table of BS4471: 1987 (see Table B1.1), such as 38 × 97 mm or 45 × 97 mm processed from 38 × 100 mm or 47 × 100 mm respectively. To comply with fire regulations, based on the available test evidence, 38 mm is the minimum thickness which can be used for stud walls required to have fire resistance.

(e) The type of preservation, if required, and method of application

It is not normal to preserve the timber of interior stud walls above the level of a screed. It is normal in the UK to preserve the timber in exterior stud walls no matter what cladding or brick skin is used as the weathering surface. NHBC require timber in exterior stud walls to be preserved. In principle, preservation should comply with BS5268: Part 5: 1989 on the preservative treatment of structural timber. To avoid the need to re-dry after preservation, it is normal to preserve timber above the level of a screed by an organic solvent. Particularly when using mixed species such as Canadian Spruce-Pine-Fir, experience shows that over-absorption of preservative can occur and can lead to problems on site with staining of plasterboard linings. To avoid this, BS5268: Part 5: 1989 (and NHBC regulations) permits the V/3 double vacuum schedule to be used 'for commercial mixed species' which can contain 'resistant' or 'extremely resistant' species, and for European Whitewood. (See Table A10.1 for preservation schedules or immersion periods which satisfy BS5268: Part 5: 1989.) Although it is normal to use an organic solvent to preserve timber above the level of the screed, CCA schedules are included in Table A10.1. When preserving construction timbers, it is normal for an organic solvent to include an

Table A10.1 Preservation of timber in exterior stud walls to satisfy BS5268: Part 5: 1989

Timber in an exterior stud wall where there is an air space behind the exterior finish	Treatment schedules or immersion periods to satisfy BS5268: Part 5: 1989 for:			
	'Permeable' or 'Moderately resistant' species 'P' or 'MR'		'Resistant' or 'Extremely resistant' species 'R' or 'ER'	
	Organic solvent	CCA water borne	Organic solvent	CCA water borne
Timber above the top level of a screed	Schedule V/1 or 10 minute immersion[2]	Schedule P2 20 g/l solution	Schedule V/4 or V/3[2,3]	Schedule P3 30 g/l solution
Timber below the level of the top of a screed[1]	–	Schedule P3 30 g/l solution	–	Schedule P4 30 g/l solution

– not permitted.
[1] Also applicable for interior stud walls.
[2] The organic solvent must include an insecticide if it is considered that there is a risk of insect attack.
[3] When treating 'commercial mixed species' and European Whitewood, the V/3 Schedule can be used by agreement.

insecticide. An insecticide is necessary if it is considered that there is a risk of insect attack.

Timber below the level of the top of a concrete screed (in an interior as well as an exterior stud wall) is said to have a slight risk of developing soft rot, therefore should not be preserved by an organic solvent (which does not give protection against soft rot). A CCA preservative should be used. It is not normally necessary for a specifier to go to the extent of detailing the actual preservation schedules which may be used. On occasions, however, the specifier may wish to state if a water borne or organic solvent preservative is to be used and, in the case of organic solvent, to know if immersion is permissible. Guidance is given, therefore, in Table A10.1 on schedules which satisfy BS5268: Part 5: 1989.

TYPICAL SPECIFICATIONS

Each specifier of the timber in a stud wall must make the final decision on what and how to specify, but the following are given as two examples of the general form of typical specifications. They do not cover all possible cases.

Typical specification for timber in an exterior stud wall with the sole plate below the level of a concrete screed.
The timber in exterior stud walls to be Canadian Spruce-Pine-Fir surfaced CLS or a European Whitewood equivalent of the same size with corners rounded not more than 3 mm, tolerances in accordance with BS4471: 1987. The sizes of studs, lintols etc. and the assembly is to be in accordance with Drawing ____. At the time of factory assembly, the moisture content is not to exceed 18%. At the time of erection, the moisture content is not to exceed 21%. With the exception of the noggings, all timber in the stud walls is to be stress graded to a level which will satisfy the Strength Class SC4 level of BS5268: Part 2: 1988. With the exception of the sole plate which is to be set below the level of the screed, all timber is to be preserved by a double vacuum organic solvent process to satisfy BS5268: Part 5: 1989. The sole plates are to be preserved with a CCA preservative complying with BS4072: Part 1: 1987 to satisfy BS5268: Part 5: 1989.

Typical specification for timber in the interior stud walls of a timber framed house with the sole plate set below the level of the concrete screed.
The timber studs and horizontal members in the interior stud walls as shown in Drawing ____ to be surfaced 38 × 63 mm with rounded corners, stress graded to match the Strength Class SC3 level of BS5268: Part 2: 1988. Tolerances on size to be in accordance with BS4471: 1987. At the time of factory assembly, the moisture content of the timber is not to exceed 18%. At the time of erection, the moisture content is not to exceed 21%. Timber above the top level of the screed is not to be preserved. The sole plates which are to be set below the top level of the screed are to be preserved with a CCA preservative complying with BS4072: Part 1: 1987, to satisfy BS5268: Part 5: 1989. At the time of fixing, the moisture content is not to exceed 21%.

A11

GLULAM MEMBERS, TIMBER FOR LAMINATES AND ROOF DECKING

INTRODUCTION

A structural glulam member (a glued-laminated timber structural member) is defined in BS4169: 1988 on the manufacture of 'glued-laminated timber structural members', as a 'member made from four or more separate laminations of timber arranged parallel to the longitudinal axis of the member, the individual pieces being assembled with their grains approximately parallel and glued together to form a member which functions as a single structural unit' (see Fig. A11.1).

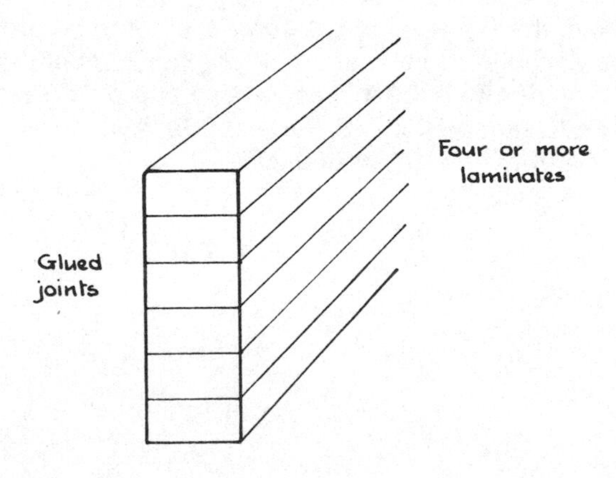

Fig. A11.1 A typical glulam section

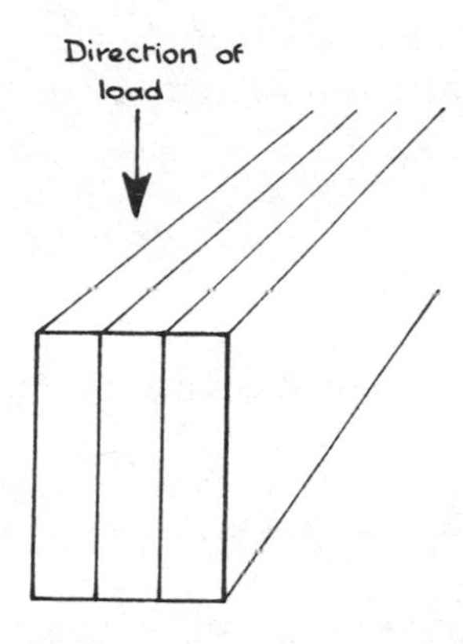

Fig. A11.2 A glued-composite rather than glulam

Members can be straight, cambered or curved, of uniform depth or tapered. The maximum finished thickness of any lamination is 45 mm.

If a joint occurs in an individual lamination, although a scarf joint is permitted, it is more normal for a finger joint of adequate strength to be used (see Chapter B10), manufactured in accordance with BS5291: 1976 on 'finger joints in structural softwood'. If a lamination contained a butt joint, the lamination would have to be excluded in the structural design.

Although BR or MR adhesives are permitted by BS4169: 1988 for certain 'exposure categories', it is normal to use a WBP gap-filling resin adhesive (see Chapter B9) in accordance with BS1204: Part 1: 1979 in the manufacture of structural glulam, and in the finger joints or scarf joints used in laminations.

If timber sections are glued together as shown in Fig. A11.2, with the load acting parallel to the glue lines, they are not considered to be glulam as defined in BS4169: 1988 even if four or more pieces are glued together, but must be designed as a glued-composite, and the stress grade of the timber must be as for solid timber, e.g. GS/MGS, SS/MSS etc. (see Chapter B3).

When glulam members are used as exposed beams or portal frames in a flat or pitched roof, it is common to support the weathering membrane on solid timber tongued and grooved decking. This forms a structural membrane and a ceiling which is normally considered to have an appearance compatible with the glulam members.

SURFACE CLASSIFICATIONS

BS4169: 1988 describes three surface classifications of completed structural glulam members which are:

'Regularized',
'Planed', or
'Sanded'.

BS4169: 1988 gives a detailed specification for each classification, but basically:

- 'Regularized' members must have at least 50% of the surfaces 'sawn or planed to remove the protruding laminations';
- 'Planed' members must have all surfaces fully planed;
- 'Sanded' members (as defined in BS4169: 1988) are 'Planed' members which are also sanded and have defects filled or otherwise made good.

The surface classification is not considered to affect the strength of the member.

STRESS GRADE OF LAMINATIONS AND MARKING OF COMPLETED MEMBERS

The strength of glulam members is determined by the stress grade of the individual laminations. These are given the designations:

LA, LB or LC.

LC is the weaker.

In future it is possible that European stress grades S6, S8 and S10 will be adopted for glulam laminates, but currently the stress grades used are LA, LB and LC as detailed in BS4978: 1988 on softwood grades for structural use. The stress and stiffness values appropriate to members made either entirely with one lamination stress grade, or a combination of two lamination stress grades are given in BS5268: Part 2: 1988. Partly because of the difficulty of providing full strength end joints for LA laminates, it is normal to base designs on the use of LB or LC laminations, or a combination of LB outer laminates and LC inner laminates.

Stress grading of laminations can be carried out visually or by machine. BS4978: 1988 requires stress grading to be controlled within a third party quality assurance scheme, and requires each completed member to be marked to identify:

- the stress grade,
- the species,
- the type of adhesive,
- the surface classification,
- the manufacturer, and
- the number of the British Standard.

If a member is preserved, a mark is also required to identify the type of preservative treatment.

'STANDARD GRADE' STOCK GLULAM MEMBERS

In addition to describing surface classifications 'Regularized', 'Planed' or 'Sanded', Appendix D of BS4169: 1988 also describes a quality of 'Standard Grade' glulam members. A high percentage of straight, uncambered glulam members manufactured from European Whitewood are imported from the Nordic area, often available from stock or on short delivery times, and it is these members which are intended to be referred to by the BS4169 definition of Standard Grade glulam members. Basically the specification for these members as described in BS4169: 1988 is:

- 'Planed' surfaces as defined in BS4169: 1988;
- 45 mm thick laminations of European Whitewood;
- LB laminations stress graded visually or by machine;
- manufactured at a moisture content of 12–15%;
- end joints in laminations of finger joints of adequate strength;
- WBP adhesive in finger joints and for assembly;
- uncambered.

If required, they can be sanded and can have defects filled or made good as required to satisfy the 'Sanded' definition of BS4169: 1988.

SOLID TIMBER DECKING

If solid timber tongued and grooved decking is used between beams, the species is normally European Whitewood or European Redwood. It is important for the designer to give clear instructions on the way in which boards are to be fixed to the supports and to each other. (Guidance is given in *The timber designers' manual* published by BSP Professional Books.) The boards are not normally stress graded, although it is usually considered acceptable to base their design on the stresses in BS5268: Part 2: 1988 appropriate to the 'Special Structural' stress grade of BS4978: 1988 and the actual species. Where the boards are exposed, the specifier may agree that there is no need for a specification for the visual appearance to be written. If one is required, however, two possibilities are to specify that the quality of 'Exposed' surfaces should match the requirements for Class 2 or Class 3 of BS1186: Part 3 on wood trim when published (see Chapters A4 and B4), or BS1186: Part 1: 1986 on the quality of timber in joinery.

It is not normally necessary to preserve decking if exposed in a building with normal humidity unless there is a local risk in the particular building. If, however, it is decided to preserve against the possibility of fungal attack rather than insect attack, possible treatments are those based on the guidance given in Table 3 of BS5268: Part 5: 1989 for the preservative treatment of structural timber, for 'Flat roofs—warm. Inverted (exposed beams)'. These are:

- for 'Permeable' and 'Moderately resistant' species, the CCA P3 20 g/l schedule or the V/1 organic solvent schedule, or
- for 'Resistant' or 'Extremely resistant' species, the CCA P7 30 g/l schedule or the V/3 or V/4 organic-solvent schedules.

It is not normal to consider exposed timber decking as being in a 'roof void', therefore there is not considered to be a risk from House Longhorn Beetles, even in an area where they are known to be active.

INDIVIDUAL LAMINATIONS

Normally it is only manufacturers of glulam, rather than specifiers or buyers of glulam members who will be involved in preparing a specification for individual laminations. The items covered will be as follows.

Species

Commonly European Whitewood is used, or European Redwood if a 'warm' appearance or extra depth of penetration of preservative is required. Douglas Fir-Larch is also used on occasions. BS4169: 1988 permits any of the species to be used for which stresses are given in BS5268: Part 2: 1988, although, if finger jointing of laminations to BS5291: 1984 is to be used, mixed species must not be used.

Stress grade

Normally LB or LC laminations are specified, or a combination of LB in the outer quarter or more of the depth of a section, with LC at the centre of the depth, visually or machine stress graded to BS4978: 1988. The limits placed on fissures in the width of individual laminations (including the exposed top and bottom widths, unless otherwise specified) are detailed in BS4978: 1988. The limits on fissures on the exposed edges of individual laminations (i.e. on the sides of completed members) are detailed in BS4169: 1988.

Size and tolerances

BS4169: 1988 limits the thickness of laminates to a maximum finished thickness of 45 mm and also defines tolerances after planing. Normally curved sections have thinner laminations. Although the sides of completed glulam members may have to be planed after manufacture, it is normal to plane individual laminations all round before assembly (i.e. not just to plane the thickness).

Moisture content

BS4169: 1988 requires the moisture content of laminations at the time of assembly to be in accordance with the requirements of the adhesive manufacturer, to be close to the service moisture content, and requires the moisture content in adjacent laminations and within the total member to be within defined close limits.

Preservation

It is not particularly normal for glulam members to be preserved unless there is a specific local hazard. The uses of glulam are so varied that, if a decision is taken to preserve, each case should be discussed with the manufacturer on how best to satisfy the requirements of BS5268: Part 5: 1989. Some of the tables in this book (e.g. Table A8.1) and the notes above related to exposed timber in flat roofs ('warm') with normal humidity may give useful guidance.

Although it is possible to preserve individual laminations before assembly of a glulam member, and although it may be necessary to do so for certain risk situations, the clear preference expressed in BS4169: 1988 is not to preserve individual laminations unless this is considered to be essential for a special case. If, however, a specifier insists on the preservation of individual laminations, it is prudent for the manufacturer to be consulted at an early stage when alternatives are being considered and costs are being estimated. The manufacturer will have to check if the chosen preservative is compatible with the assembly adhesive and process, and with the finger jointing process.

If using a water borne preservative, the manufacturer will also have to decide if preservation is to be carried out before the laminations are re-dried and planed (which would remove much of the preservative and require the waste to be disposed of in an environmentally safe manner), or to be preserved after the laminations are planed. If the intention is to preserve after the laminations have been finger jointed, a check must be made to see if sufficiently long preserving equipment is available. Also, an adequate time must be left between assembling the joints and preserving, to ensure full cure of the adhesive before the preservation process is begun.

If an organic solvent is to be used to preserve individual laminations, a formulation containing a water repellant is not permitted by BS4169: 1988.

Applying a flame retardent

On occasions, there may be a requirement to apply a flame retardent (see Chapter B6). In such a case, it is essential that the manufacturers of the flame retardent and the adhesive are consulted before the specification is finalised, to ensure that the flame retardent and the timing of application are compatible with the adhesive and with the service environment of the glulam member.

TYPICAL SPECIFICATION OF INDIVIDUAL LAMINATIONS

As indicated above, a typical specification for individual laminations is normally best left to the manufacturer to ensure that the final product complies with BS4169: 1988. Within a manufacturing company, however, the in-company specification could take the following form:

> 'Laminations to be European Whitewood LB stress grade of BS4978: 1988; moisture content to be an average of 14% with no reading exceeding 17%; planed after finger jointing to 45 × 122 mm to the tolerances of BS4169: 1988.

SPECIFYING COMPLETED GLULAM MEMBERS

The specification of a completed glulam member is often evolved in discussion between the building designer, the manufacturer and the structural designer. (Often the structural designer works for the manufacturer.) Particularly if preservation or the application of a surface spread of flame retardant is required, it is prudent for discussion to take place at an early stage with the intended manufacturer. If preservation is required, a species must be chosen which can be preserved to the required level in accordance with BS5268: Part 5: 1989. Before specifying a pressure or double vacuum process, it is essential to check if sufficiently large equipment is available, to check the suitability of the process for the species, and the compatibility with the adhesive.

The specification should cover the following items:

Species, stress grade, thickness of laminations, and moisture content

The notes given above for specifying individual laminations apply.

Surface classification

Either 'Regularized', 'Planed', 'Sanded' or 'BS4169: 1988 Standard Grade glulam member'. See the notes above.

Size of members, tolerances and camber

Often the depth of uncambered or cambered members will be in increments of 45 mm (because of the normal thickness of laminations), although this is not essential because glued 'make-up' pieces can be added either top or bottom. If members are 'Planed' or 'Sanded', typical finished widths are as listed below.

The maximum deviations on the cross-section of completed sections are given in BS4169: 1988 as:

On finished widths of members up to 220 mm	± 3 mm
On wider sections	by agreement
On finished depths of members up to 300 mm	± 3 mm
from 301 mm to 600 mm	± 6 mm
from 601 mm to 900 mm	± 10 mm
over 900 mm	± 12 mm

Typical finished widths of 'Planed' members to which the size tolerances of BS4169: 1988 given above apply are:

Finished width (mm)	Depth of sections (mm)
65 or 66	180 to 495
90	180 to 720
115	180 to 900
135 or 140	180 to 900
160	180 to 900
185 or 190	180 to 900
215	180 to 900

The width of 'Sanded' members is likely to be slightly less.

Very wide members can be manufactured from two widths of laminations (see Fig. A11.3). Certain limitations apply (not detailed here).

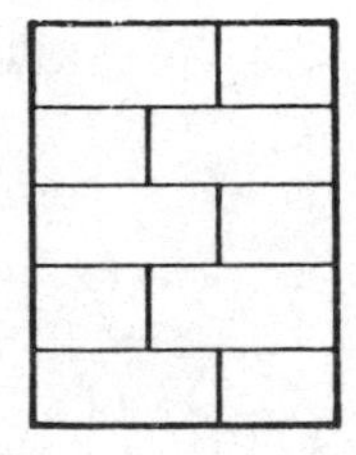

Fig. A11.3 Joints in the width of glulam laminates

If the design calls for a width in excess of about 190 mm or a depth in excess of about 900 mm, it is prudent to discuss the weight, handling etc. with the manufacturer before finalizing the building details.

The structural designer should agree with the manufacturer whether or not a camber is to be provided. Stock members are normally manufactured without a camber.

Manufacturing standard and adhesive

Normally the members should be manufactured in accordance with BS4169: 1988. Normally the adhesive used in the assembly process and in the finger joints should be a WBP gap-filling resin adhesive complying with BS1204: Part 1: 1979. BS4169: 1988 includes requirements for the protection of members during storage and erection. It is quite common for members, particularly imported members, to be wrapped with a protective paper or film during transit.

End joints within individual laminations

Although scarf joints are permitted, end joints in laminations are normally finger joints (see Chapter B10).

Preservation

See notes above including those related to individual laminations.

TYPICAL SPECIFICATION OF A COMPLETED GLULAM MEMBER

There are many possible ways of specifying glulam members and the wording will vary depending on whether the specification is by a manufacturer with design staff or by a specifier. A typical specification could take the form indicated below:

> Glulam members are to be the sizes shown on Drawing ______. The members are to be manufactured and marked in accordance with BS4169: 1988 including complying with the tolerances on sizes. At the time of manufacture, the moisture content is to be suitable for a service moisture content of 14%. The surface classification is to be 'Planed' as defined in BS4169: 1988. Completed members are to be provided with a protective covering during transit.
>
> Members are not to be cambered.
>
> Individual laminations are to be European Whitewood, stress graded LB in accordance with BS4978: 1988, and having a finished thickness of not more than 45 mm. End joints in laminations are to be finger joints with adequate strength for the design.
>
> The adhesive used in the finger joints and in the assembly of the members is to be a gap-filling WBP adhesive in accordance with BS1204: Part 1: 1979.

TIMBER FOR PLY-WEB BEAMS

A12

INTRODUCTION

Ply-web beams are less popular than they were when a large number of schools were being built with flat roofs in the UK at the time of the raising of the school leaving age (ROSLA), but they are still manufactured and used, mainly to support roofs, normally in situations where there is a ceiling underneath the beams. Unless specially manufactured or finished, ply-web beams are not normally considered to have an appearance suitable for being left exposed in a dwelling, school etc.

A ply-web beam can be used as a single member, for example as a purlin, but it is more normal to use a system of beams at centres not normally in excess of 1.2 m. Members are manufactured in depths of up to about 1.2 m although the most popular depths are in the order of 600–800 mm. Typical cross-sections of I and Box beams are shown in Fig. A12.1, and a typical cross-section through a roof is shown in Fig. A12.2.

Although some beams are assembled using only nails, the more normal construction is to use an adhesive between the timber flanges and stiffeners and the plywood web or webs. In most designs, nails are used to hold the joints together during curing of the adhesive. The nails are then left in place, although not considered in the calculations of the

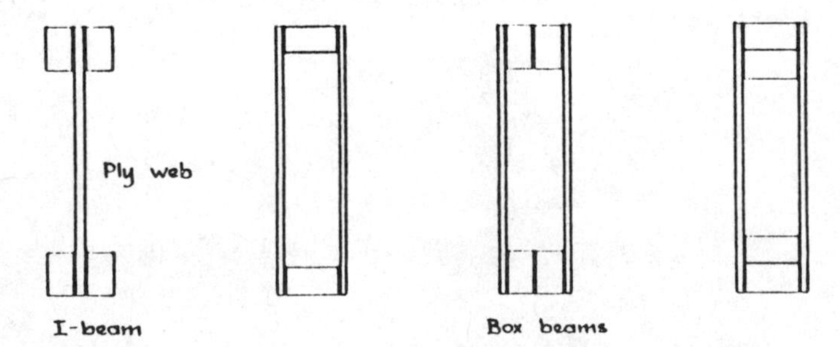

Fig. A12.1 Typical cross-sections of ply-web beams

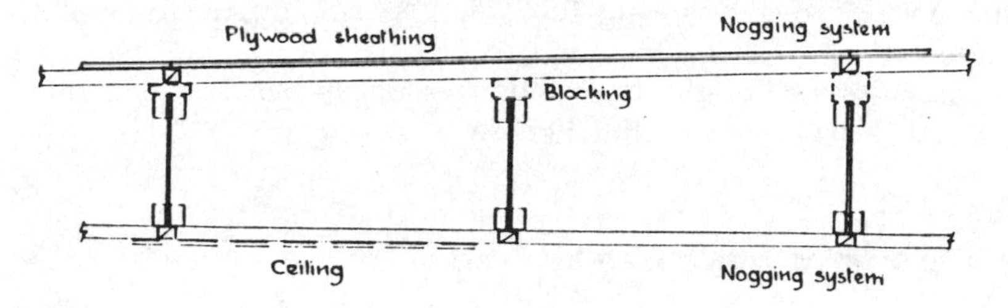

Fig. A12.2 Typical cross-section through a flat roof constructed from ply-web beams

strength of the joints. The use of glued joints enables full composite action to be assumed in the structural design. For ply-web beams assembled with adhesive, manufacture should be in accordance with BS6446: 1984 on 'the manufacture of glued structural components of timber and wood based panel products'.

Assemblies such as ply-web beams are considered in BS6446: 1984 not to be suitable for use in 'Exterior high hazard' situations (see Table B9.2).

For a structural glued assembly, the surfaces to be glued must be processed (planed) a short time before being glued, and therefore sawn timber complying with BS4471: 1987 is normally chosen by the manufacturer and planed within 24 hours before assembly of the members. The flange members must be stress graded, normally to BS4978: 1988 on softwood grades for structural use, and therefore each piece must have a finished size of not less than 35 × 60 mm. When flange members are stress graded to BS4978: 1988, it is not normally considered necessary to mark every piece but to accept that one mark per beam is adequate to identify the stress grade, species etc. (The marking requirements of BS6446: 1984 do not cover this point.)

Joints in the length of the flanges can be scarf joints but structural finger joints in accordance with BS5291: 1984 are more normal.

Except in the special cases of designs with curved webs, web stiffeners are normally required. They are not normally fully stress graded unless required to transmit a specific concentrated load, but are planed and free from wane, and glued into place.

It is normal to manufacture ply-web beams, particularly long span ply-web beams with a camber, although one proprietary ply-web beam manufactured by a continuous process with a curved web is not cambered.

In the past it has been normal for the noggings between beams (see Fig. A12.2) to be planed from basic sizes of 50 × 50 mm (or even smaller at the ceiling line), and not to be stress graded. In future, it is likely that stress grading will be more strictly enforced, including stress grading of the noggings supporting the roof decking, in which case the minimum size to comply with BS4978: 1988 would have to be 35 × 60 mm.

A specification for the timber flanges in ply-web beams should include details of Items (a) to (f) below.

(a) The species which is to be used

The most common species for the flanges of ply-web beams is European Whitewood, with European Redwood being used if preservation by immersion is required. BS6446: 1984 permits any of the species to be used for which strength properties are given in structural design code BS5268: Part 2: 1988. However, if finger jointing is to be used, mixed species are excluded by BS5291: 1984. Also, it is necessary to choose a species which will be available suitably dried, and a species available in suitable sizes.

(b) The stress grade which is to be used

Because of the unsuitability of certain species (as indicated above), it is safer to specify the species which are acceptable as well as the stress grade (or Strength Class) rather than only specify a Strength Class (see Chapter B3).

For the flanges it is normal to specify one of the stress grades of BS4978: 1988. Experience shows that the permissible stresses of 'General Structural' are on the low side for ply-web beams, and the choice of which stress grade to specify lies between 'Special Structural' (SS or MSS), M50 or M75. Special Structural can be stress graded visually as well as by machine. This fact gives a certain flexibility of supply which can decide the specifier to select SS/MSS. Wane is permitted by BS4978: 1988 up to the grade limit but should be excluded in a specification for timber for ply-web beams.

The timber for the web stiffeners is not normally stress graded (unless short ends of stress graded timber are used). One exception may be the use of stress graded timber for loadbearing stiffeners.

(c) The moisture content at the time of manufacture

BS6446: 1984 requires the moisture content of the timber (and the plywood) at the time of planing and assembly to be suitable for the glued joints between the timber and the plywood, and the glued joints in the length of the flanges. Timber must be within 5 percentage points of the service moisture content (see Table B2.2).

(d) The size of the flanges, type of processing, and tolerances on size

The smallest size currently permitted to be stress graded to BS4978: 1988 is 35 × 60 mm. On the other hand, BS6446: 1984 states that 'The finished size of each piece at right-angles to a glueline shall not exceed 50 mm.' By reference to the right hand example sketched in Fig. A12.1, it can be seen that a pedantic interpretation of the BS6446 clause would limit some flanges to 50 × 50 mm (and hence not satisfy the 60 mm minimum requirement of BS4978). The intention, however, of the BS6446 clause in this case is that, where a face (rather than an edge) is glued, the thickness at right angles to that face should not exceed 50 mm. (Note that the BS6446 clause should not be applied to the glue line between the plywood and the timber joists in a stress-skin panel because it is the edges, not the faces, which are glued.) Finished sizes of pieces used in the flanges of a ply-web beam could be:

35 × 60 mm, 35 × 70 mm, 44 × 70 mm, 44 × 95 mm etc.

Note that, in the case of the flanges shown in the right hand example in Fig. A12.1, the larger dimension is planed twice, once before the two pieces in each flange are glued together, and then again to ensure that both pieces are in line before being glued to the plywood.

Tolerances after planing should be to BS4471: 1987.

The tolerances permitted for completed beams should be in accordance with BS6446: 1984.

(e) The adhesive to be used

Although BS6446: 1984 permits several adhesives to be used depending on the service condition (see Table B9.2), it is normal to use a gap-filling WBP resin adhesive complying with BS1204: Part 1: 1979 both for assembly of flanges to webs and in the finger joints, even though assemblies such as ply-web beams are not considered by BS6446: 1984 to be suitable for use in 'Exterior: high hazard' situations.

(f) Preservation, if required

Because of the many different cross-sections possible with ply-web beams and the many situations in which ply-web sections are used, it is very difficult to give general guidance on preservation other than to advise caution. If the specifier decides that ply-web beams must be preserved for a specific risk such as House Longhorn Beetles in a 'roof void', the manufacturer should be consulted before the specification and the building design is finalized. Preservation should be in accordance with BS5268: Part 5: 1989 on the preservative treatment of structural timber. The tables in this book may give relevant guidance.

It may be possible to preserve the timber pieces with a water borne preservative before the timber is re-dried, planed and assembled, but planing will remove much of the preservative. The preservative and the adhesive must be compatible.

The timber pieces may be preserved with an organic solvent after finger jointing and planing but before assembly, but the preservative and the adhesive must be compatible.

It may be possible to treat completed beams if equipment of sufficient size is available, and if adequate time is left between gluing and preserving. If, however, a pressure or vacuum treatment is used, the effect of the pressure or vacuum on the completed assembly and the integrity of the plywood must be considered. At the very least, it may be found necessary to add additional web stiffeners. It is necessary also to consider the possibility of preservative being trapped within the profile of Box or other beams which have two or more webs.

TYPICAL SPECIFICATION FOR THE TIMBER FLANGES OF A PLY-WEB BEAM

A typical specification for flanges of a ply-web beam could be:

'Flanges to be as shown in Drawing _____, planed to 44 × 70 mm with tolerances in accordance with BS4471: 1987. Species to be European Whitewood stress graded SS or MSS to BS4978: 1988. Wane is not permitted. At the time of planing and assembly, the moisture content is to be between 14% and 18%.'

A13 LARGE SOLID TIMBER ENGINEERING SECTIONS

INTRODUCTION

Rather than write a chapter on specifying timber for cases where large solid sections (i.e. not laminated or built up from several smaller sections) are required for timber engineering, the object of these notes is to draw attention to possible pitfalls such as the risk of specifying sizes or grades which are unlikely to be available, or which may prove to be unsuitable in practice.

SIZE, WIDTH AND THICKNESS

Although it is possible to obtain European timber in sizes up to 300 mm wide and in thicknesses of 100 mm, it is not easy to find quantities of European timber larger than 75 × 225 mm or 75 × 250 mm. Sizes of 75 × 250 mm and over are more likely to be available in North American timbers, and it is always prudent to check on supply before finalizing a design which is based on large pieces of solid timber. See Chapter C5 for the sizes and species of Canadian timber most likely to be available in large sizes.

MOISTURE CONTENT

It can be difficult to kiln dry large sizes and, to air dry large sizes in the UK can be a lengthy process.

The UK structural design code BS5268: Part 2: 1988 calls for solid timber thicknesses of over 100 mm to be designed normally using 'wet' stresses and stiffness values unless the timber is specially dried.

Most large sizes available in the UK are of Canadian timber exported undried to the UK (see Chapter C5).

GRADE, DISTORTION AND JOINTS

It is more difficult to dry thicknesses of over 75 mm without the timber developing fissures and end splits during drying. This point should be discussed with potential suppliers, particularly if the performance or appearance of a component such as a roof truss, or the strength of a connection depends on there being limited or no fissures in the timber. (Often it is preferable to use two or more thinner sections joined together rather than one thick piece.)

When designing a triangulated framework or any item which requires large sections of timber to be joined together, it should be remembered that even small distortions or cupping in large sections can make it difficult for the section to be pulled into line which, for example, can affect joints, particularly glued joints, and can add secondary stresses to the timber and joints.

If finger jointing is necessary to achieve a long length of a large section, a check should be made to ensure that a finger jointing plant with sufficient capacity to handle the section is available. Many plants are not able to handle sizes wider than 250 mm or thicker than 75 mm (or are not set up to do so).

The stress grade can be one of the NLGA grades or a stress grade from BS4978: 1988 on softwood grades for structural use (see Chapter B3). It should be noted, however, that the largest size which can be machine stress graded by machines currently used in Europe is 75 × 300 mm, therefore only visual stress grading is applicable for larger sections.

PRESERVATION

If the specification calls for the timber to be preserved, particularly if the thickness is in excess of 75 mm, the extent to which preservation will be effective, and any special requirements to make it effective should be discussed with the supplier, manufacturer or a specialist.

TIMBER FOR TILING OR SLATING BATTENS ON PITCHED ROOFS OR VERTICAL WALLS A14

INTRODUCTION

The sizes, tolerances, species and grade of the tiling or slating battens used in the design cases shown in Table 3 of BS5534: Part 1 (see Table A14.1), are covered in detail in BS code BS5534: Part 1: 1978 (1985) on *Slating and tiling. Design*, including amendments AMD3554: 1982 and AMD5781: 1988. If larger or longer span battens are used, they must be designed in accordance with structural design code BS5268: Part 2: 1988, using one of the stress grades which satisfy BS5268: Part 2: 1988.

A specification for the timber for tiling or slating battens in the sizes covered by Table 3 of BS5534: Part 1: 1978 (1985) should include details of Items (a) to (e) below.

(a) The species which is to be used

Details of the species which is to be used or reference to the species permitted by BS5534: Part 1: 1978 (1985) to satisfy the sizes quoted in Table 3 should be given. These species, as defined in normal trade terms rather than as in BS5534: Part 1; are:

European: Redwood or Whitewood;
Canadian: Spruce-Pine-Fir, Douglas Fir-Larch, or Hem-Fir;
British-grown: Sitka Spruce or Scots Pine.

European Redwood and European Whitewood are the most common species. There is some usage of British-grown and other timbers.

(b) The grade which is to be used

Grade limits are detailed in amendment AMD3554 of BS5534: Part 1

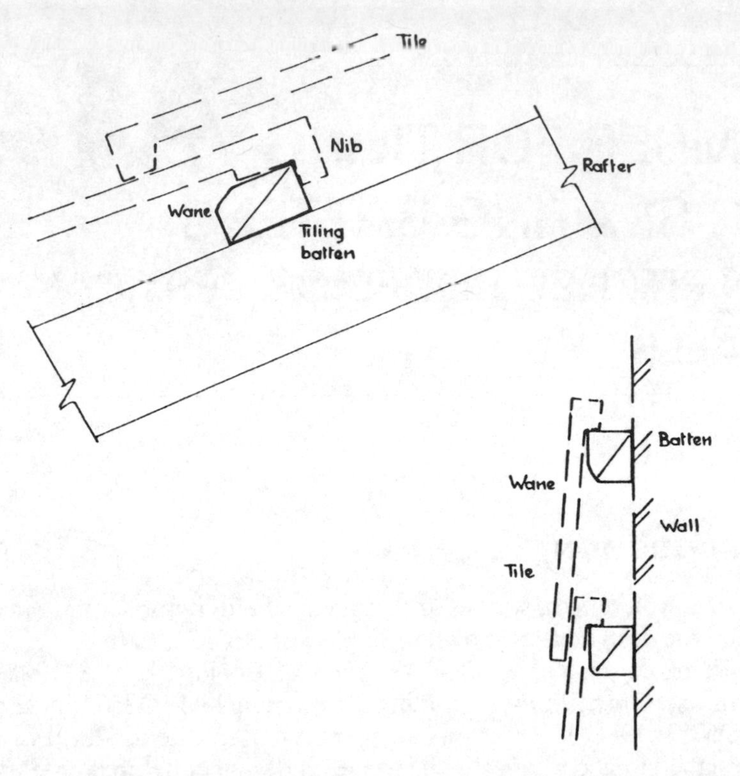

Fig. A14.1 Preferred position of wane if it occurs in a tiling or slating batten

by cross-reference to clauses in the 1973 version of BS4978 which relate in principle to visually stress graded 'General Structural'. Despite this, there is no intention or requirement for tiling or slating battens to be stress graded or to be graded only by authorized graders, or for the timber to be grade marked. Wane is permitted up to one third of the dimension of the surface on which it occurs. The preference is that, if wane occurs, it should be positioned in a pitched roof or on a vertical wall as sketched in Fig. A14.1.

(c) The moisture content at time of erection

Clause 11.1 of BS5534: Part 1: 1978 (1985) makes the rather unusual comment that 'there is no need to limit closely the moisture content because drying in situ is not restricted', but then goes on to call for the moisture content not to exceed 22%. Although not specifically intended for tiling battens, the most relevant reference in BS5268: Part 2: 1988 permits a moisture content of up to 24% at the time of erection.

A large percentage of battens to support roof tiles or slates are preserved, as are nearly all battens to support vertical tiles. In these cases, normally the preservative is a water borne CCA which dramatically increases the moisture content of small sections. It is not usual to kiln dry battens after preservation, therefore time should be allowed for air drying before they are delivered and fixed.

(d) The size, type of surface, and tolerances on size

Table 3 of BS5534: Part 1: 1978 (1985), as amended by AMD3554 and AMD5781 gives the sizes and tolerances of battens for pitched roofs and vertical work to support a variety of slates and tiles. Battens must be continuous over at least two spans. Details of sizes and permissible tolerances are given in Table A14.1 with part of the explanatory notes. A note in AMD3554 about the Explanatory Memorandum, Part 3, linked to the Building Standards for Scotland is likely to be out of date

Table A14.1 Recommended batten sizes for pitched roofs and vertical work from BS5534: Part 1: 1978 (1985)

Examples of slating or tiling	Basic sizes[1]			
	450 mm span		600 mm span	
	Width (mm)	Depth (mm)	Width (mm)	Depth (mm)
Slates				
Sized	38	19	38	25
Random	50	25	50	25
Asbestos-cement	38	19	38	25
Concrete	38	19	38	25
Clay and concrete tiles				
Plain: pitched roofs	32	19	32	25
vertical work	38	19	38	25
Single lap	38	22	38	25

[1] All the sizes in this table are subject to production/re-sawing allowances of *either* width +3 mm, −2 mm *or* depth +3 mm, −1 mm. In addition, the sizes are subject to a minus tolerance of 0.5 mm in width *and* depth.[2] (Note that 'width' is the larger dimension.)

Where other batten sizes are used, they should be not less in both width and depth than those given in the table. To avoid splitting, the maximum nail diameter should not exceed one-tenth of the batten width for nails fixed on the centre line of the width (face) of the batten. For other cases, the size of the batten should be designed in accordance with BS5268: Part 2: 1988.

[2] Experience has shown that there can be confusion about how the tolerance clauses linked to the sizes in AMD3554 and AMD5781 are intended to be applied, therefore it is advantageous to consider an example. Note that, although tiling and slating battens are normally produced by re-sawing, the tolerances are slightly more generous than those permitted for resawn sizes in BS4471: 1987 on the sizes of sawn and processed softwood. If one takes the case of a batten having a basic sawn size (or 'resawn ex-larger'—see Chapter B1) of 38 × 19 mm, the minimum size can be either 35.5 × 18.5 mm or 37.5 × 17.5 mm. The maximum size permitted in this case is 41 × 22 mm.

BS5534: Part 1: 1978 (1985) and BS5268: Part 3: 1985 require battens to be at least 1200 mm long and to be supported at least at three positions.

so is not repeated with Table A14.1. The specifier should take note, however, of the latest guidance on the practice in Scotland of the use of various sarking materials under battens on sloping roofs.

Battens for use on pitched roofs and vertical work normally have a sawn surface. Battens are normally resawn from wider sections.

(e) The type of preservation, if required, and method of application

Battens in 'dry' pitched roofs

As can be seen from reference to Table B5.4, and considering the risk of fungal attack on the outside of the sarking of a normal 'dry' pitched roof as being 'low' or 'negligible' (i.e. Risk category 1 or 2), and the risk of insect attack as being equally low, except where there is known to be a particular local risk, preservation of tiling or slating battens on pitched roofs is normally 'unnecessary' or 'optional'. In recent years, however, it has become normal practice in the UK to preserve a high percentage of battens. Currently over half of battens in pitched roofs are preserved.

It should be understood that, even in an area where House Longhorn Beetles are active, tiling battens normally fixed outside sarking are not within the warm 'roof void' as intended for the Approved Document to support Regulation 7 of the Building Regulations 1985 for England and Wales, and hence do not need to be preserved against House Longhorn Beetles.

Although an organic solvent can be used, if preservation is carried out, it is normal to use the cheaper water borne CCA process, and to call for one which complies with BS4072: Part 1: 1987.

Table A14.2 Preservation, if required, of battens in pitched roofs

Battens in a 'dry' pitched roof[1]	Treatment schedules or immersion periods to satisfy BS5268: Part 5: 1989 for:			
	'Permeable' or 'Moderately resistant' species 'P' or 'MR'		'Resistant' or 'Extremely resistant' species 'R' or 'ER'	
	Organic solvent	CCA water borne	Organic solvent	CCA water borne
	Schedule V/1 or 3 minute immersion[2]	Schedule P8 15 g/l solution	Schedule V/3 or V/4 [2,3]	Schedule P9 15 g/l solution

[1] See the text in this chapter on 'risk'.

[2] The organic solvent must include an insecticide where it is considered that there is a risk of insect attack.

[3] When treating 'Resistant' or 'Extremely resistant' species by an organic solvent against fungal attack, immersion is not permitted for tiling battens.

Because tiling and slating battens are small, BS5268: Part 5: 1989 on the preservative treatment of structural timber, permits special short CCA Schedules P8 or P9 as relevant to the amenability of species to preservation (see Table A14.2), which were developed to prevent over-absorption with small sizes. It is possible to misinterpret Table 4 of BS5268: Part 5: 1989, therefore it is worth stating that the intention is that Schedules P8 and P9 may be used to preserve tiling battens on dry pitched roofs if one is preserving against both decay and insect attack, as well as if preserving against either.

Battens to support vertical tiles

Although Building Regulations are not specific on the need or otherwise to preserve battens to support vertical tiles, NHBC regulations call for them to be preserved, although the level of preservation is not defined. The thinking behind the call for preservation is that wind and slight building faults or damage can lead to rain or snow penetrating to the battens. In normal well constructed and maintained work, there is only a 'low risk' of fungal attack and even less risk of insect attack unless there is a particular local hazard.

An organic solvent or water borne preservative can be used.

Although Table 4 of BS5268: Part 5: 1989 is clearly the correct source for selecting the schedules to use when preserving tiling or slating battens for pitched roofs, it is not clear if this is the best source for selecting the schedules to use when preserving battens to support vertical tile hanging, timber cladding etc. Another possible reference source is Table 5 ('external woodwork in buildings above the damp-proof course') of code BS5589: 1989 on the preservation of timber, even though the primary intention of this table is to give guidance on the preservation of exposed, decorated and maintained timber in windows etc.

In the absence of specific guidance, the author is inclined to suggest reference to Table 4 of BS5268: Part 5, particularly because this refers to the use of the short CCA Schedules P8 and P9 as relevant for small sections. It should be noted that, even in an area where House Longhorn Beetles are active, there is no need to preserve against them in a vertical wall, only in a 'roof void'.

TYPICAL SPECIFICATIONS

Each specifier of tiling or slating battens must make the final decision on what and how to specify, but the following are given as three examples of the general form of typical specifications. They do not cover all possible cases.

Typical specification for battens in a 'dry' pitched roof with trussed rafters at 600 mm centres where it is decided to preserve the battens.

Tiling battens to be one of the species which comply with BS5534: Part 1: 1978 (1985) *Slating and Tiling. Design*, of basic sawn size 38 × 25 mm with tolerances complying with AMD3554 and AMD5781, preserved to BS5268: Part 5: 1989 with a CCA preservative complying with BS4072: Part 1: 1987.

No individual length is to be shorter than 1200 mm.

The moisture content as supplied to site is not to exceed 22%.

Typical specification for tiling battens in a 'dry' pitched roof with rafters at 450 mm centres where there is no requirement for the battens to be preserved.

Tiling battens to be European Whitewood or European Redwood of basic sawn size 32 × 19 mm complying with BS5534: Part 1: 1978 (1985) and amendments. The moisture content as supplied to site is not to exceed 22%.

Typical specification for battens to support vertical tiles on an exterior wall of a dwelling.

Battens to be basic sawn size 38 × 25 mm complying with BS5534: Part 1: 1978 (1985) and amendments, preserved with a CCA preservative complying with BS4072: Part 1: 1987 with Schedule P8 (15 g/l) of BS5268: Part 5: 1989 if a 'Moderately resistant' species is used or Schedule P9 (15 g/l) if a 'Resistant' species is used.

The moisture content as supplied to site is not to exceed 22%.

A15

TIMBER FOR EXTERIOR CLADDING, FASCIAS, BARGE BOARDS AND SIMILAR EXTERIOR TRIM

INTRODUCTION

Until recently there had been understandable confusion about the grading rule or British Standard to refer to when specifying the quality of timber in exterior cladding, fascias etc. One tendency was to refer to a sawmill grade such as 'fifth quality' or 'unsorted' when specifying European timber but, as explained in Chapter B4, these grading rules are not precise, are not end use grades, and are invalidated if re-sawing takes place. As an alternative, some specifiers referred to one of the Classes of BS1186: Part 1 on the quality of timber in joinery, but this standard is over-complex for a specification for components such as cladding and fascias. The 1986 version emphasized this point by stating that BS1186: Part 1 does not apply to cladding, profiled boards or wood trim.

Schedule 14 of the Building Standards (Scotland) Regulations 1981 still refers to sawmill grades (which are not end use grades—see Chapter B4), but the situation on specifying exterior cladding and exterior wood trim has been clarified by the drafting of BS1186: Part 3 *Timber for and workmanship in joinery. Specification for wood trim: timber species, classification, workmanship in fixing, and standard profiles*, which is expected to be published in 1990. It is intended as a simplified version of BS1186: Part 1: 1986 and, as drafted, it also includes a few clauses on fixing, finishing and standard profiles. It is intended to be used as the basis for specifications of exterior barge boards, fascia boards, architraves, weather trims, exterior timber cladding and ancillary trim, interior skirting boards, architraves etc., and interior timber panelling and ancillary trim (see Chapter A4). When BS1186: Part 3 is published, the intention is to withdraw BS584: 1967 (1980) *Specification for wood trim (softwood).*

In the following notes, the assumption is made that BS1186: Part 3 will be published generally as drafted at the time of 'public comment' in 1988 and as cleared, in 1989, for publishing by BSI.

EXTERIOR CLADDING

In the UK, it has been common for exterior cladding to be planed and to be assembled from horizontal tongued and grooved boarding. It is not the intention that this book should cover design, but it is relevant to the drafting of specifications to note that experience in the Nordic area and in the UK has shown that the best performance of timber cladding is obtained with relatively thick square edged sections, and with a sawn surface on the outside.

Boards can be fixed horizontally (overlapped), but better performance is achieved in general with boards fixed vertically i.e. batten-on-board, board-on-board, or board-on-batten (see Fig. A15.1). Also, it is preferable to 'turn' boards and battens so that any reduction in moisture content which causes the boards to cup will tend to tighten the joints between them (see the growth ring configuration in Fig. A15.1, and see Fig. B2.1). The nailing to horizontal support battens should be arranged to allow free movement of each board or batten (see Fig. A15.1).

The first choice of species in the Nordic area (from the choice of European Redwood or European Whitewood) is European Whitewood, which has a natural property of resisting the ingress of

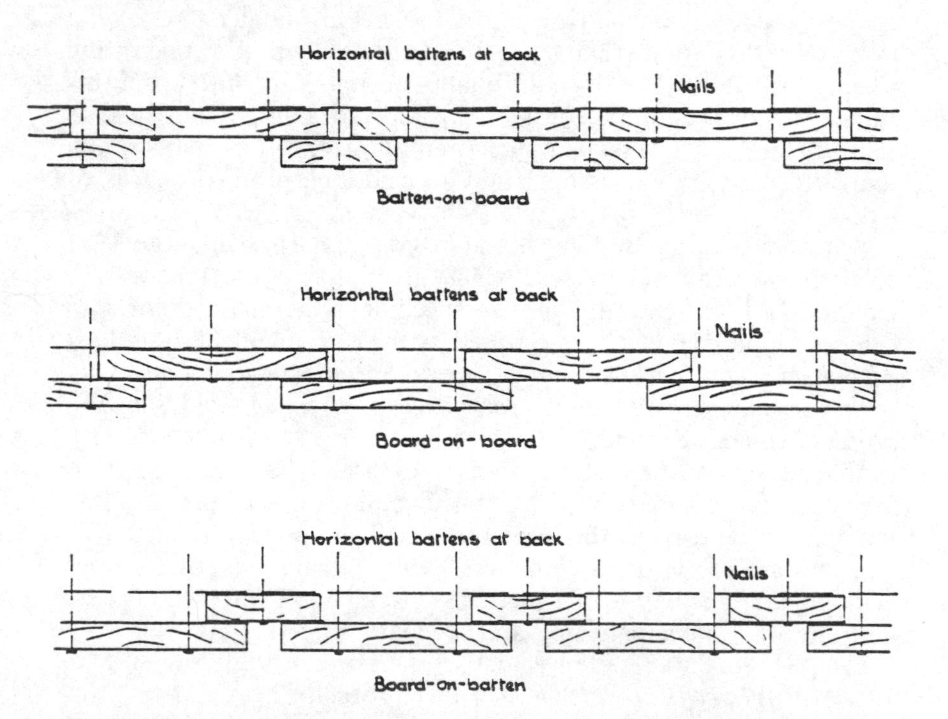

Fig. A15.1 Typical cross-sections through vertical square-edged cladding boards

moisture from rain to a remarkable extent. This property enables a relatively thin exterior-quality stain or relatively thin coat of an exterior-quality opaque finish (such as an exterior-quality acrylic emulsion) to be used and to perform well (see Chapter B7), because the timber itself provides much of the required resistance to rain.

FASCIAS AND BARGE BOARDS

Fascias and barge boards are usually planed.

SPECIFICATIONS

A specification for timber in exterior cladding, fascias, barge boards etc. should include details of the following Items (a) to (e), which may vary in the form of the wording depending whether it is written by a specifier or a builder.

(a) The species which is to be used or a list of the species which are acceptable or could be used

The most common species for exterior cladding, fascias etc. are European Whitewood or European Redwood. Of these two species, as stated above, European Whitewood has the advantage for exterior use of having a natural property of resisting the ingress of moisture to a remarkable extent.

For special effect, Canadian Western Red Cedar is used.

Although suitable for exterior uses, Canadian Hemlock or Canadian Douglas Fir-Larch are not often used for exterior cladding, fascias etc. in the UK.

(b) The Class or Classes of timber required in the final work as fixed

If European Whitewood or European Redwood are specified, the most relevant Classes of BS1186: Part 3 (as drafted) are Classes 2 and 3. If a stain or other non-opaque finish is to be used, the specifier should consider calling for shakes and checks to be left un-filled, and should consider specifying that the exterior surfaces of exterior cladding should be left with a sawn surface rather than a planed surface.

If Western Red Cedar, Canadian Hemlock or Canadian Douglas Fir-Larch are specified, the most relevant Classes of BS1186: Part 3 are Classes 1 and CSH. The comments made above in relation to

European timbers about considering the use of a sawn surface apply also to these species. (The shakes and checks permitted in Classes 1 and CSH are too narrow to be filled.) According to BS1186: Part 3 as drafted, Western Red Cedar and Douglas Fir-Larch are 'Susceptible to iron staining in damp conditions', therefore the use of non-ferrous fasteners is advisable in exterior conditions.

(c) The moisture content at the time of delivery and fixing

BS1186: Part 3 as drafted indicates that 'exterior wood trim' as defined is likely to have a moisture content in service of between 13% and 19%. It is desirable for cladding, fascias etc. to have a moisture content within this range at the time of delivery and fixing. (Note that, unlike BS1186: Part 1: 1986, Part 3 does not give average values of moisture content.)

A high percentage of timber supplied for exterior cladding and wood trim will be preserved with a water borne preservative which affects the moisture content readings of a moisture meter (see Chapter B2). Timber preserved by a water borne process should be dried before being fixed.

(d) Sizes, tolerances and surface quality

BS1186: Part 3 as drafted includes details of certain standard profiles of:

tongued and grooved boards,
shiplap,
weatherboards,
skirting boards,
architraves,
half-rounds, cover-fillets, quadrants and scotia mouldings,

which are intended to be useful to specifiers. Standard codings are given to aid cross-referencing, particularly if a drawing is not available.

Cladding

To limit movement in service, it is better to limit the width of cladding boards to basic dimensions of 100 mm or 125 mm if tongued and grooved or 'shiplap' (see Fig. A15.2), with 150 mm being regarded as the maximum sensible width if square-edged boards or 'weather-boards' are used. 'Deemed-to-satisfy' example G9 in Schedule 13 of the

Building Standards (Scotland) Regulations 1981 limits the width of vertical tongued and grooved boards to 100 mm and the width of horizontally fixed boards to 150 mm.

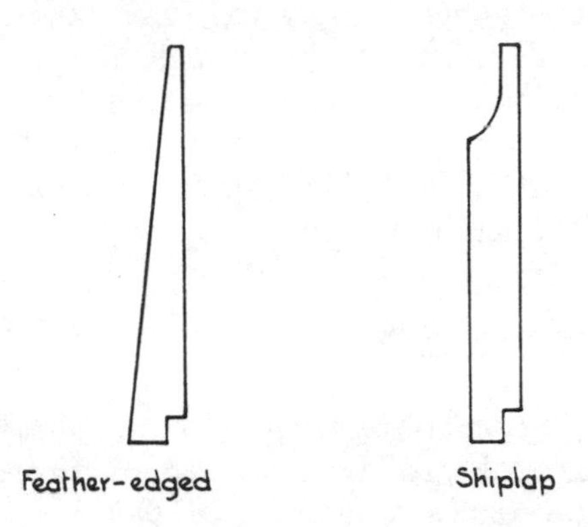

Fig. A15.2 Typical cross-sections through shiplap and feather-edged boards

The minimum thickness recommended by the Building Regulations 1985 for England and Wales is 16 mm finished. In the case of feather-edged boards (see Fig. A15.2), the thickness of 16 mm applies at the thicker edge, with the thinner edge dimension being at least 6 mm. 'Deemed-to-satisfy' example G9 in Schedule 13 of the Scottish Regulations calls for boards to be not less than 21 mm thick. In the case of tapered boards, the dimension of 21 mm applies at the thicker edge. In the Nordic area, experience in practice has shown that there are functional advantages in using boards of at least 19 mm finished thickness.

The width of the tongues of tongued and grooved boards, and the overlap of other weatherboards should be adequate to take account of movements in service.

The tolerances of BS4471: 1987 on sawn and processed softwood apply (i.e. no minus tolerance permitted on processed sizes measured at 20%). See Chapter B1 for further relevant information on BS4471: 1987.

As stated earlier in this chapter, there are advantages for performance, including the performance of stains and low-build opaque finishes such as exterior-quality acrylic emulsions, in using a sawn rather than a planed surface on the outside of cladding boards. If the surface is planed, it should comply with BS1186: Part 3 as drafted (i.e. there should be between 8 and 18 cutter marks per 25 mm).

Fascias, barge boards etc.

Fascias to support gutters will be dimensioned to suit individual architectural details but, on domestic property, a typical basic depth is 200 mm planed to about 190 mm (see Tables B1.3 and B1.4 in Chapter B1). The same comments apply to barge boards. Typical finished thicknesses are 19 mm or 22 mm which are possible to achieve by planing the common 'Whitewood thickness' of 22 mm sawn or the common 'Redwood thickness' of 25 mm sawn. Tolerances should be in accordance with BS4471: 1987, surface quality in accordance with BS1186: Part 3 (when published).

Exterior mouldings or trim

Exterior mouldings will be dimensioned to suit individual architectural details. The face quality can be sawn or planed as required. Tolerances should be in accordance with BS4471: 1987.

(e) Preservation, if required

With the exception of Canadian Western Red Cedar and Canadian Douglas Fir-Larch, BS1186: Part 3 (as drafted) and the National House-Builders Council (NHBC) require exterior cladding, fascias and barge boards to be preserved in accordance with BS5589: 1989 which is the code of practice for preservation of timber. NHBC also requires battens supporting cladding to be preserved (see Chapter A5).

Neither the Building Regulations 1985 for England and Wales nor the Building Standards (Scotland) Regulations 1981 are precise in stating that timber cladding must be preserved. Clause 1.4 of the Building Regulations 1985 and Regulation B2(ii) of the Scottish Regulations 1981 may be read as permitting preservation to be considered to be 'unnecessary' where cladding is readily accessible for inspection and maintenance or renewal but, if preservation is carried out, it is required to be in accordance with BS5589. To satisfy the Building Regulations for England and Wales, it is not normally necessary to preserve Douglas Fir-Larch or Western Red Cedar. The Scottish Regulations do not require Western Red Cedar to be preserved, but do not exclude Canadian Douglas Fir-Larch from the list of species required to be preserved.

Fascias and barge boards in two storey buildings and higher are hardly easy to maintain or renew, and a probable interpretation of the Regulations is that they should be preserved.

In general, the Building Regulations (Northern Ireland) 1977 will be satisfied if the relevant British Standards and Codes of Practice are satisfied. The Table to Regulation B3 and Tables 2 and 3 call for the

Table A15.1 Preservation processes which can satisfy BS5589: 1989 for exterior cladding and wood trim

Species and amenability of the heartwood to preservation[1]	Desired service life performance category and preservation process					
	30 years			60 years		
	Organic solvent		CCA	Organic solvent		CCA
	Double vacuum	Immersion		Double vacuum	Immersion	
European Whitewood 'R'	+	–	+	–	–	+
European Redwood 20 mm thick or less 'MR'	+	+ (3 min)	+	+	+ (60 min)	+
thicker than 20 mm 'MR'	+	+ (3 min)	+	+	–	+
Canadian Hemlock 'R'	+	–	+	–	–	+
Canadian Douglas Fir-Larch 'R'	+	–	+	–	–	+

+, acceptable; –, not permitted.
[1] i.e. Resistant 'R' or Moderately Resistant 'MR' (see Table B5.3).

softwoods covered by this book, except Western Red Cedar, to be preserved if used as exterior timber boarding (cladding).

When satisfying BS5589: 1989, the relevant schedules are detailed in Table 4 of BS5589: 1989. It is necessary for the specifier to decide on a desired service life performance category. The choice is between 30 or 60 years. Normally the 30 year performance category should be adequate for preservation of cladding, fascias etc.

If the decision is taken to preserve to satisfy BS5589: 1989, Table A15.1 details the preservation methods which can be used. As a general guide, if a moulded section is to be preserved, an organic solvent preservative is likely to be favoured, and if a sawn section is to be preserved, the (cheaper) water borne CCA process may be favoured. For planed fascias and barge boards which are to be planed, the specifier has the choice of a CCA or organic solvent process and may well favour an organic solvent applied after planing. See Chapter B5 for further reading.

DISTANCE OF TIMBER CLADDING FROM A BOUNDARY

If consideration is being given to the use of timber cladding on a wall close to a boundary, a check should be made on the rules in the relevant Building Regulations on the permitted proximity of combustible materials to a boundary, and the rules for 'unprotected areas'. They are too detailed to explain here. A few notes are given in Chapter B6.

BATTENS TO SUPPORT EXTERIOR CLADDING

For notes on the specification of battens to support exterior cladding, see Chapter A5.

TYPICAL SPECIFICATIONS

Each builder or specifier of timber must make the final decision on what and how to specify, but the following are given as four examples of the typical content of specifications for cladding and exterior wood trim as defined in BS1186: Part 3 (as drafted).

Typical specification for vertical sawn timber cladding, board-on-board, in England.
The cladding boards shown on Drawing ____ are to be European Whitewood of basic sawn size 22 × 150 mm to match Class 3 of BS1186: Part 3: 19__, preserved with a CCA preservative complying with BS4072: Part 1: 1987 to the 30 year desired service life performance category of BS5589: 1989. At the time of fixing, the moisture content is not to exceed 19%. The exterior sawn surfaces are not to be 'made good'. (See Item ____ for the specification and colour of the exterior quality acrylic emulsion which is to be applied.)

Typical specification for vertical tongued and grooved cladding boards and associated trim in England.
The tongued and grooved cladding boards and associated trim detailed on Drawing ____ are to be European Whitewood to Class 2 of BS1186: Part 3: 19__. The T&G boards are to be moulded from basic sawn size 22 × 100 mm with an exterior planed surface complying with BS1186: Part 3. Making good is to be carried out in accordance with BS1186: Part 3 prior to the sections being finished with an acrylic emulsion as specified in Item ____. All timber is to be preserved with an organic solvent to the 30 year desired service life performance category of BS5589: 1989. At the time of delivery to site and fixing, the moisture content is not to exceed 19%.

Typical specification for fascia and barge boards in domestic property in England.
Fascia and barge boards are to be European Whitewood planed to 19 × 190 mm in accordance with BS1186: Part 3: 19__ from 22 × 200 mm, graded to Class 3 of BS1186: Part 3. Preservation to the 30 year desired service life performance category of BS5589: 1989 is to be by a water borne CCA preservative complying with BS4072: Part 1: 1987, or an organic solvent. At the time of fixing, moisture content is not to exceed 19%. The exterior surface is to be made good in accordance with BS1186: Part 3 for an opaque exterior quality solvent based paint.

Typical specification for horizontal shiplap boarding in Scotland.
Shiplap boards moulded to the profile shown in Drawing ____ are to be European Whitewood or European Redwood to Class 3 of BS1186: Part 3: 19__ preserved by an organic solvent to the 30 year desired service life performance category of BS5589: 1989. The moisture content is to comply with BS1186: Part 3. The surface quality is to comply with the planing requirements of BS1186: Part 3. No making good of fissures etc. is to be carried out.

A16 TIMBER FOR SCAFFOLD BOARDS AND TOEBOARDS

INTRODUCTION

It is not often that a specifier is called on to prepare a detailed specification for scaffold boards, therefore this chapter differs from the main chapters in this section of the book, in that it gives background notes in Items (a) to (e) below which may be useful if a specifier is called on to check scaffold boards, or if someone is buying or selling scaffold boards.

The relevant British Standard for timber scaffold boards is BS2482: 1981, which describes scaffold boards of basic sizes 38 × 225 mm, 50 × 225 mm and 63 × 225 mm. The National Association of Scaffolding Contractors (NASC) has also produced a specification for scaffold boards, the latest version having been printed in 1987. Although it refers to BS2482, the grading and marking differ somewhat from those in BS2482, and refer only to a basic size of 38 × 225 mm.

The relevant code of practice is BS5973: 1981 for access and working scaffolds and special scaffold structures of steel.

The only mention of 'toeboards' found in British specifications is in BS5973: 1981 which calls for them to 'have a minimum height of 152.5 mm'. One presumes that this is a metric conversion from 6 inches and that a 150 mm piece of sawn timber would be acceptable if fixed above the scaffold boards, although common practice is to use a scaffold board on edge, with the lower edge at the level of the underside of the horizontal scaffold boards.

(a) The species

The species permitted to match BS2482: 1981 and the NASC Specification are listed in Table A16.1. For conformity with BS2482: 1981, some species are included in Table A16.1 which are not otherwise dealt with in this book. European Whitewood is the most common species used for scaffold boards and toeboards.

Table A16.1 Species for timber scaffold boards

Species in BS2482: 1981 and the NASC Specification 1987
Imported European Whitewood
Imported European Redwood
Canadian Douglas Fir-Larch
Hem-Fir (Hemlock)
Spruce-Pine-Fir
British-grown European Larch
Japanese Larch
Douglas Fir[1]
Scots Pine

[1] Suitable for machine graded boards only.

(b) The grade and marking

BS2482: 1981 describes a visual grade and also gives settings for machine stress grading of scaffold boards. With visual grading, the limits for the size of knots are based on the way in which knots appear on a surface, not on the 'knot area ratio' (KAR) method as described in BS4978: 1988 on softwood grades for structural use. The grade descriptions in the NASC Specification differ somewhat from those in BS2482: 1981 and apply only to the size 38 × 225 mm.

To date, only a small percentage of scaffold boards have been machine stress graded. BS2482: 1981 does not insist on machine stress grading being carried out within the 'Kitemark' scheme. To do so would require the machine operator to be in a different scheme to that applicable to BS4978: 1988. Some boards have been machine stress graded to appropriate limits without compliance with BS2482 being claimed.

BS2482: 1981 requires the following marking:

- The number of the British Standard.
- The identification mark of the supplier.
- The letter M or V to denote machine or visual grading.
- The word 'support' followed by the maximum span in metres over

Table A16.2 Sizes and spans of scaffold boards

Sizes and maximum spans in BS2482: 1981 and BS5973: 1981			
Basic size (mm)	**Minimum size (mm)**	**Maximum size (mm)**	**Maximum span (m)**
38 × 225	36 × 220	40 × 230	1.5
50 × 225	47 × 220	53 × 230	2.6
63 × 225	60 × 220	66 × 230	3.25

Sizes are measured as at 20% moisture content.

which the board can be supported, followed by the abbreviation 'max' (see Table A16.2 for the maximum spans permitted by BS2482: 1981).

The NASC Specification requires the following marking:

- The name or identification mark of the supplier.
- The words 'support at 1.5m max span'.

Both BS2482: 1981 and the NASC Specification stipulate how end hoops or nail plates are to be fixed.

Neither BS2482: 1981 nor BS5973: 1981 refer to a special grade for toeboards. One presumes that the only requirement is for toeboards to be sufficiently straight to be fixed in place, and to be free from decay and live insect attack. (Often scaffold boards 'on edge' are used for toeboards.)

(c) The moisture content

Neither BS2482: 1981 nor the NASC Specification 1987 require the timber to be at a specific moisture content when delivered or in service, although BS2482: 1981 calls for sizes and tolerances to be measured as at 20% moisture content. (One advantage of using European Whitewood for scaffold boards is that it is a species which resists the ingress of moisture from rain to a remarkable extent hence, once dried, will not change moisture content as readily as other species.)

(d) The size, type of surface, and tolerances on size

BS2482: 1981 lists three basic sizes of scaffold boards with maximum spans over which they can be used in scaffolding. Amendment AMD 4657: 1984 states that 'The face surface shall be sawn; edges shall be sawn or planed.' The sizes and maximum spans are given in Table A16.2 with the permitted tolerances indicated by 'minimum' and 'maximum' sizes, all measured as at 20% moisture content, whether the edges are sawn or planed. The tolerances and maximum span for the 38 × 225 mm basic size board agree with the requirements of the NASC Specification 1987.

BS5973: 1981 calls for boards of 38 mm basic thickness, 3.9 m long to be supported on at least four transomes, with boards shorter than 3.35 m to be supported on at least three transomes. One presumes that boards with a length of between 3.9 m and 3.35 m should be supported on at least four transomes. BS5973: 1981 does not permit boards shorter than 2.13 m to be used unless they are fixed down to prevent tipping.

Neither BS2482: 1981 nor BS5973: 1981 define a minimum thickness for toeboards, but BS5973: 1981 requires the top of toeboards to be at least 152.5 mm above the top of the scaffold boards. One presumes, however, that a 150 mm deep sawn board would be considered to be adequate (say 22 mm thick). Common practice, however, is to use a scaffold board on edge as a toeboard, with the lower edge at the level of the underside of the horizontal scaffold boards.

(e) Preservation, or resistance to surface spread of flame

It is not normal for scaffold boards for normal building purposes to be preserved, nor to be chemically treated to increase the resistance to the surface spread of flame. For special uses, however, such as one could encounter on oil rigs, such treatments may be required.

TIMBER FOR SPECIAL FENCES

A17

INTRODUCTION

A great deal of timber is used in the UK in the manufacture of fences and gates. Most fencing components are sold 'as seen' or by sample or brochure with no detailed specification being written, although often the species is stated. There are cases, however, such as Ministry of Transport contracts or purchases of high class fences, where a written specification is regarded as being necessary. The main object of this chapter is to draw attention to a series of Part British Standards on fences under the general number of BS1722, and to indicate the detail in them. They are:

BS1722: Fences:

- Part 1: 1986 *Specification for chain link fences.*
- Part 2: 1973 *Woven wire fences.*
- Part 3: 1973 *Specification for strained wire fences.*
- Part 4: 1986 *Specification for cleft chestnut pale fences.*
- Part 5: 1986 *Specification for close boarded fences.*
- Part 6: 1986 *Specification for wooden palisade fences.*
- Part 7: 1986 *Specification for wooden post and rail fences.*
- Part 8: 1978 *Mild steel continuous bar fences.*
- Part 9: 1979 *Mild steel fences with round or square verticals and flat posts and horizontals.*
- Part 10: 1972 *Anti-intruder chain link fences.*
- Part 11: 1986 *Specification for woven wood and lap boarded panel fences.*
- Part 12: 1979 *Steel palisade fences.*
- Part 13: 1978 *Chain link fences for tennis court surrounds.*

There is particular reference to timber in Parts 2, 4, 5, 6, 7 and 11. The notes below indicate the extent of the detail included in Parts 5, 6, 7 and 11.

For details of preservation refer to Section Six of BS5589: 1989 *Code of practice for preservation of timber.*

TIMBER FENCES

BS1722: Part 5: 1986 *Close boarded fences*

Although Appendix B of Part 5 permits the use of any softwood (or hardwood), in other respects it is very detailed. It gives precise details of the size and tolerances of posts, arris and rectangular main rails, capping and counter rails, feather edged board filling, gravel boards, centre stumps and timber cleats, and it tables grading limits for posts, main rails and 'non-structural components'. It calls for preservation, when required by the choice of species, to be in accordance with BS5589, and suggests that the 20 year desired service life performance category (Category B) is normally appropriate for most cases rather than the 40 year category (Category A).

BS1722: Part 6: 1986 *Wooden palisade fences*

Although Appendix B of Part 6 permits the use of any softwood (or hardwood), in other respects it is very detailed. It gives precise details of the size and tolerances of posts, arris rails and rectangular rails, palisades and centre stumps, and it tables grading limits for posts, rails and 'non-structural components'. It calls for preservation, when required by the choice of species, to be in accordance with BS5589, and suggests that the 20 year desired service life performance category (Category B) is normally appropriate rather than the 40 year category (Category A).

BS1722: Part 7: 1986 *Wooden post and rail fences*

Although Appendix A of Part 7 permits the use of any softwood (or hardwood), in other respects it is very detailed. It gives precise details of the size and tolerances of main posts, prick posts and rails for morticed and nailed fences, and tables grading limits for main posts, prick posts and rails. It calls for preservation, when required by the choice of species, to be in accordance with BS5589, and suggests that the 20 year desired service life performance category (Category B) is normally appropriate rather than the 40 year category (Category A).

BS1722: Part 11: 1986 *Woven wood and lap boarded panel fences*

Although Appendix B of Part 11 permits the use of any softwood (or hardwood) for the posts, and any softwood for the timber for panels, in other respects it is very detailed. It gives precise details of the size and tolerances of posts, vertical and horizontal battens, gravel boards, vertical stiffeners, horizontal rails, cappings, slats and boards, and tables grading limits for posts, post caps and fillets, and panels. It calls for preservation, when required by the choice of species, to be in accordance with BS5589, and suggests that the 20 year desired service life performance category (Category B) is normally appropriate rather than the 40 year category (Category A).

SPECIFYING

When specifying the timber to satisfy one of the Parts of BS1722, the choice of species, and the level and type of preservation can be left to the manufacturer to satisfy BS1722, or can be selected by the specifier or purchaser. Other than these two points, provided that there is a Part of BS1722 to cover the type of fence which is required, it seems adequate simply to base the other parts of the specification on reference to the relevant Part of BS1722.

B1 SAWN SIZES AND PROCESSING OF SIZES

SIZES AND TOLERANCES

All the significant European softwood importing and exporting countries changed softwood sizes to metric measure in 1971. (In this context, European includes the Nordic countries.) BS4471: Part 1: 1971 *Dimensions for softwood. Sizes of sawn and planed timber* and BS4471: Part 2: 1971 *Dimensions for softwood. Small resawn sections* were published. Part 1 was updated in 1978 and was rewritten in 1987. Part 2 on small resawn sections was found not to be useful and was withdrawn in 1987, therefore BS4471: Part 1 became simply BS4471: 1987 *Specification for sizes of sawn and processed softwood.*

Unfortunately for standardization of sizes used in the UK, it was not possible for the Canadian sawmilling industry to adopt metric measure at the same time as Europe but, around 1985, some Canadian mills began to saw to the metric sizes and tolerances of BS4471 for part of their export to the UK.

Although the primary object of the 1971 version of BS4471 was to present a table of agreed 'basic' metric sawn sizes, of equal importance was the decision to eliminate the previous permissible minus tolerance of $\frac{1}{8}$ inch. Small minus tolerances ('maximum permitted minus deviations' as defined in BS4471: 1987) are still permitted by BS4471 on basic sawn sizes, but only on 10% of pieces. The agreement to virtually eliminate minus tolerances was made possible by the increased accuracy of sawing and drying at European sawmills.

The 1971 and 1978 versions of BS4471 permitted small minus tolerances on processed sizes, but the 1987 version eliminated permissible minus tolerance on dimensions between opposed processed surfaces.

The basic sizes tabulated in BS4471: 1987 are measured at 20% moisture content. The 'basic' size is defined as the size to which tolerances apply. Table B1.1 shows the basic sawn sizes given in BS4471: 1987. Chapters C1 to C8 of this book give the variations which are applicable for ranges of sizes from various supply countries. For example, although a certain amount of 250, 275 and 300 mm wide and

Table B1.1 Basic sizes of sawn softwood as presented in BS4471: 1987

Basic thickness (mm)	Basic width (mm)								
	75	100	125	150	175	200	225	250	300
16	×	×	×	×					
19	×	×	×	×					
22	×	×	×	×	×	×	×		
25	×	×	×	×	×	×	×	×	×
32	×	×	×	×	×	×	×	×	×
36	×	×	×	×					
38	×	×	×	×	×	×	×	×	×
44	×	×	×	×	×	×	×	×	×
47	×	×	×	×	×	×	×	×	×
50	×	×	×	×	×	×	×	×	×
63		×	×	×	×	×	×		
75		×	×	×	×	×	×	×	×
100		×		×		×	×	×	×
150				×		×			×
200						×			
250								×	
300									×

Notes:
1. Normally 47 mm thick sizes are sawn for construction rather than joinery uses.
2. From the Nordic area, 22 mm thick sizes are usually available mainly in Whitewood, and 25 mm thick sizes mainly in Redwood.
3. 44 mm thick sizes in BS4471 widths are not readily available in the UK.
4. Widths of 115 mm and 275 mm are sometimes available.
5. Although a certain amount of 250, 275 and 300 mm wide, and 100 × 100 mm timber is available from European (including Nordic) mills, the main volume of European timber is in sizes of 75 × 225 mm and smaller (see Chapters C1–C4, C6 and C8). By and large, thicker or wider sizes are of North American origin and to NLGA sizes and tolerances, not BS4471 sizes and tolerances.

100 × 100 mm sawn sizes are available from European mills, the main volume is in sizes of 75 × 225 mm or smaller. To a large extent, thicker or wider sizes are of Canadian origin in NLGA (National Lumber Grades Authority) or *Export R List Grading and Dressing Rules* Imperial sizes and tolerances, not BS4471 sizes and tolerances.

The maximum permitted deviations in BS4471: 1987 on sawn sizes not exceeding 100 mm are:

minus 1 mm, plus 3 mm

and, on sawn sizes of over 100 mm are:

minus 2 mm, plus 6 mm.

Minus tolerances are permitted on only 10% of pieces in any ‘parcel’. A ‘parcel’ is defined as a ‘quantity of sawn timber of the same basic size, quality and description’, but the minimum number of pieces in a

Table B1.2 CLS surfaced sizes as given in Appendix A of BS4471: 1987

Basic thickness (mm)	Basic width (mm)						
38	63	89	114	140	184	235	285

Notes (also see Tables C5.1 and C5.2):
1. CLS surfaced softwood has rounded arrises not exceeding 3 mm radius.
2. No minus tolerance is permitted (at 19% moisture content). Oversize is not limited.

parcel is not defined. (Presumably a 'parcel' can also refer to a quantity of processed timber.)

Basic sizes are measured at 20% moisture content. For any higher moisture content up to 30%, to comply with the main body of BS4471: 1987, the size of timber must be greater by 1% for every 5 percentage points of moisture content in excess of 20% (but see the notes below for timber of North American origin on the effect on size of drying). Similarly, for any moisture content lower than 20%, notes in the main body of BS4471: 1987 permit the size to be 1% smaller for every 5 percentage point reductions in moisture content.

Although some Canadian mills are sawing an increasing percentage of stress graded sawn timber to BS4471: 1987 sizes and tolerances and produce 'surfaced' Canadian Lumber Standard (CLS) stress graded timber which has metric sizes quoted in BS4471: 1987 (see Table B1.2) on which no minus tolerance is permitted at 19% moisture content, much of the Canadian timber for construction and nearly all which is intended for joinery is in Imperial sizes with tolerances which comply with NLGA rules (see Chapter C5). The permitted minus and plus tolerances on dimensions for some grades are significantly larger than the BS4471 tolerances. Canadian sawn timber is often referred to within the UK timber trade as a 'nominal' sawn size, which refers to the size, normally before drying, to which the NLGA tolerances apply, and the size on which the cost per volume unit is based. When the term 'nominal' size is used in relation to CLS sizes (see Tables C5.1 and C5.2), it is not the size to which tolerances apply, but is the size on which cost per volume unit is normally based. The term 'nominal' is deprecated by BSI although it is still used regularly within the UK timber trade.

EFFECT ON SIZE OF DRYING TO LOW MOISTURE CONTENTS

As stated above, BS4471: 1987 permits sizes dried to levels below 20% to be slightly smaller than the minimum size permitted at 20% moisture content. Many sawmills, importers etc. are able to dry timber to moisture contents as low as 10%, 12%, 15% etc. for specific uses.

When calling for sawn or processed timber at a low moisture content, the specifier should realize that the timber is normally permitted to be smaller than the basic sawn size or the minimum processed size which is permitted at 20% moisture content. The difference will be quite small but, for certain uses, can be an important consideration. For example, a basic sawn size of European timber 50 × 100 mm dried to 10% moisture content is permitted to be 49 × 98 mm, *less* the BS4471 minus tolerances on 10% of pieces. Normally it will be costed as the full basic size.

Note that, for softwoods of North American origin, Appendix A of BS4471: 1987 quotes a 1% reduction or increase in size for every 4 percentage points of change in moisture content rather than the 5 percentage points quoted in the main body of the standard.

BASIC LENGTHS

The basic lengths of sawn softwood quoted in BS4471: 1987 in metres are:

1.8	2.1	3.0	4.2	5.1	6.0	7.2
	2.4	3.3	4.5	5.4	6.3	
	2.7	3.6	4.8	5.7	6.6	
		3.9			6.9	

No minus tolerance is permitted. Overlength is not limited.

Unless produced by finger jointing, lengths of 6.0 m and over will generally be available only in North American species.

North American timber is normally sold in even 2 feet lengths, although some is sold in 1 foot increments (see Chapter C5).

RE-SAWING

It is important to realize that, when a piece of timber is resawn, the saw will usually remove 2.5–3.0 mm of timber; thus a piece which was, for example, 50 mm thick will not be a full 25 mm when resawn into two pieces. BS4471: 1987 permits up to 2 mm reduction on the basic size due to re-sawing, therefore a size produced by re-sawing 50 mm into two could be 23 mm and still satisfy BS4471: 1987 (see Fig. B1.1).

BS4471: 1987 permits a basic resawn size, for example, of 38 × 50 mm to be:

38 × 48 mm, or 36 × 50 mm,
or even 36 × 48 mm (if resawn on both axes).

The seller of resawn timber to BS4471: 1987 is supposed to refer to it either by the actual size or as the basic resawn size with the words 'resawn ex larger' added. This, however, is not always possible in practice because, on many occasions, a specification is written long before the timber is supplied, at a time when the supplier may not know if the timber will be as sawn from the log or produced by re-sawing. Although 2 mm is not particularly large, if size is important to within a millimetre, the specifier should be aware of this point and specify accordingly, especially when dealing with smaller and thinner sizes.

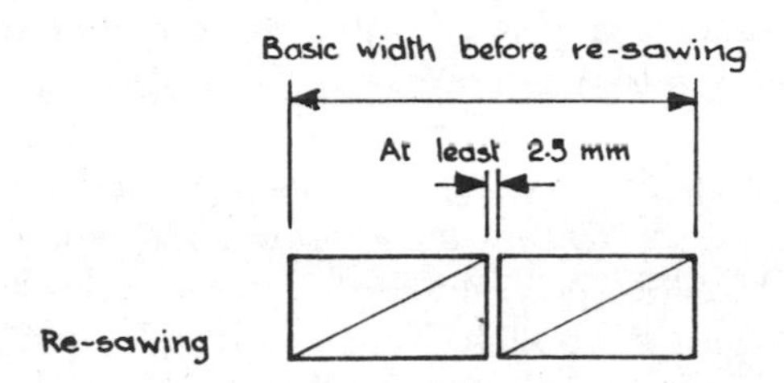

Fig. B1.1 The effect of re-sawing on size

A list of small resawn sizes used to be given in Part 2 of BS4471 but was found to be so incomplete compared to the variety of sizes being supplied that Part 2 of BS4471 was withdrawn in 1987. Clauses which cover the permitted tolerances on resawn sizes are now included in BS4471: 1987. As described above, a re-sawing reduction of 2 mm is permitted. This is not additional to any reduction due to planing or other processing, but is additional to any allowance for the timber being at a low moisture content. Thus, for example, if two pieces of timber resawn from a 75 mm thickness at 20% moisture content are required to be processed to a finished thickness of 35 mm, it is likely that the process will be able to remove only 1 mm of thickness from each piece (or less if 35 mm has to apply at say 10% moisture content).

A table of the sizes of timber normally used for slating and tiling battens is given in AMD3554 of BS5534: Part 1: 1978 (1984) (see Table A14.1 of this book). Although these are invariably produced by re-sawing, the minus tolerances permitted differ slightly from the minus tolerances permitted in BS4471: 1987 for resawn sections.

PROCESSING SAWN TIMBER

It is common to refer to processed timber by its actual size (with a drawing where necessary, for example, for a moulded section) and/or, in the case of rectangular sections, by the size from which it was processed (e.g. ex 38 × 125 mm, or 35 × 119 mm planed from 38 × 125 mm). If the price is quoted on a volume basis it is normally related to the size *before* processing. Mouldings are normally priced per unit length.

There are several methods of processing. As well as re-sawing as described above, three common processes are planing, moulding and 'regularizing' (as defined in BS4471: 1987). In addition, Canadian CLS sizes (i.e. Canadian Lumber Standards sizes) are 'surfaced', and some Canadian timber is referred to as being 'dressed' or 'dimensioned'. Regularizing and planing are described below. The Canadian terms 'surfaced', 'dressed' or 'dimensioned' may be regarded as processes to remove part of the surface of sawn timber to make the thickness and width uniform, but not to the same extent as would be associated with planed joinery or planed wood trim. The sawn sizes from which Canadian timber is surfaced, dressed or dimensioned are not defined.

Planing, and moulding to a profile are processes which are generally understood. BS4471: 1987 gives maximum permitted reductions from basic sizes of sawn softwood to finished size by planing on two opposed faces. These are shown in Table B1.3. For joinery in particular, and for wood trim, however, the maximum reductions given are not indicative of common practice, therefore values are given

Table B1.3 Maximum permitted reductions from basic sizes of sawn softwood to finished size by planing of two opposed faces as given in BS4471: 1987

	Basic sizes (mm)			
	15 to 35 incl.	Over 35 to 100 incl.	Over 100 to 150 incl.	Over 150
	Reductions from basic sizes (mm)[1]			
Constructional timber	3	3	5	5
Matching[2] and interlocking boards[3]	4	4	6	6
Wood trim other than that specified in BS584[4]	5	7	7	9
Joinery and cabinet work[5]	7	9	11	13

[1] This is inclusive of any re-sawing allowance.
[2] The reduction in width is overall the extreme size and is exclusive of any reduction of the face by the machining of a tongue or lap joint.
[3] Not applicable to flooring (see BS1297: 1987).
[4] The intention is that BS584 will be replaced by BS1186: Part 3 (when published)—see Chapter B4.
[5] See Table B1.4 for values typical of those used by industry, rather than these maximum values.

Table B1.4 Typical reductions from basic sizes to finished size of two opposed faces when planing or moulding joinery sections or wood trim

	Basic sizes (mm)				
	15–50	63	75	100–150	Over 150
	Typical reductions (mm)[1]				
Joinery and wood trim	3–6	5–7	6–8	6–9	By special consideration of end use

[1] The reductions for softwood tongued and grooved flooring are given in BS1297: 1987 (see Chapter A6).

in Table B1.4 as agreed with industry representatives as being typical of the reductions used when planing and moulding.

The tolerances given in BS4471 for processing used to be given on a plus or minus basis but, in the 1987 version, were changed not to permit any minus deviation. (The term 'permitted deviation' is used in BS4471: 1987 rather than 'tolerance'.)

Normally the minimum quoted size of a processed section is the size which would apply at 20% moisture content (or at 19% moisture content in the case of Canadian CLS sizes) even if the section is supplied at a different moisture content. Note, however, that in the case of normal domestic tongued and grooved softwood flooring complying with BS1297: 1987 (see Chapter A6), the minimum permitted size applies at the specified moisture content.

Regularizing or surfacing

'Regularizing' as defined in BS4471: 1987 is a process by means of which every piece of a batch of constructional timber is sawn and/or

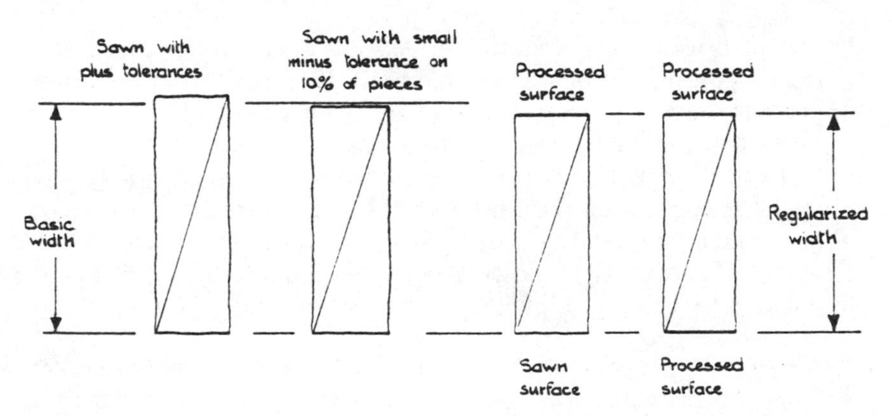

Fig. B1.2 Regularizing as intended by BS4471: 1987

machined to a uniform width. For example, in the case of joists supporting a ceiling, it may be desirable to make the depths of sawn timber more regular. To regularize, sections may be passed between two saws or cutters, or one saw or cutter and a straight edge. Thus, as shown in Fig. B1.2, one or both edges may be processed.

The reductions for regularizing given in BS4471: 1987 are 3 mm for dimensions up to and including 150 mm, and 5 mm for larger dimensions. The manufacturing deviations are plus 2 mm minus 0 mm.

CLS sizes are said to be 'surfaced' but there is no note of the amount of timber removed when surfacing. The sizes quoted in BS4471: 1987 for CLS (e.g. 38 × 89 mm) are finished dry sizes (at 19% moisture content) on which no minus tolerance is permitted. Oversize is not limited.

Standard regularizing is stated in BS4471: 1987 as removing 3 mm or 5 mm from the width depending on size (see above). There is no reason, however, why a specifier or manufacturer should not use the principle of this process but vary it, with the agreement of all concerned, particularly if making the thickness (as well as the width, or instead of the width) of constructional timber more uniform.

Wane

It should be realized that, even after planing, regularizing or surfacing, it is permissible for wane to be present up to the grade limit of constructional (or joinery) timber, unless specifically excluded by the specification.

Costing processed sizes

When the price of processed timber is quoted on a volume basis it is normal for the price to be worked out on the basis of the basic sawn size or the basic size 'resawn ex larger' from which the section has been processed. If the buyer is in any doubt as to the basis on which the timber will be costed, a check should be made to eliminate the chance of later dispute. Normally the price will be based on the basic sawn size but the buyer should pay particular attention if the words 'ex' or 'nominal' are used. The following examples indicate the type of price differences which can be involved if there is misunderstanding.

- If a piece of planed timber finishing 35 × 119 mm dry has been taken from a piece with basic size of 38 × 125 mm and is costed on a volume basis related to the 38 × 125 mm size, the buyer would calculate a price 12% too-low if assuming that the invoice would be calculated on 35 × 119 mm at the quoted rate per m^3.

- If a piece of sawn timber is quoted as 47 × 200 mm 'green' (i.e. undried), if the minus tolerances permit it to be as small as 43 × 192 mm when dry, the buyer would calculate a price 14% too-low if assuming that the invoice would be calculated on the basis of 43 × 192 mm at the quoted rate per m^3.
- It is common for CLS to have the price quoted per 'Standard' (i.e. a Petrograd Standard which is equal to 4.671 m^3), and for sizes to be charged as though processed from a nominal size of the next larger Imperial size. For example, a dry metric size of 38 × 89 mm is often costed as though processed from 2 inch × 4 inch (50.8 × 101.6 mm). The buyer would calculate a price 53% too-low if assuming that the invoice would be calculated on 38 × 89 mm at a rate quoted on the basis of a price per 'Standard'.

Obviously, in all such cases, there is less chance of confusion if the timber is costed per metre length.

DRYING AND MOISTURE CONTENT

B2

INTRODUCTION

After a tree is felled, in the case of softwood, the moisture content of the sapwood can be as high as 120–180%. (Readings as high as 250% have been encountered in Western Red Cedar.) The moisture content of heartwood is generally lower at 40–100%. Most quality sawn softwood should be dried to around 18–22% moisture content or lower before being graded, manufactured or used.

Softwood becomes stronger as it dries (unless subjected to very high temperatures for a considerable time). It becomes less prone to decay as it dries, and even sapwood is relatively immune to decay at moisture contents of 20/22% or lower. Drying can be carried out at the sawmill or at a later stage by an importer, merchant or manufacturer. Where a low moisture content is required, sometimes a first drying (to about 18%) is carried out at the sawmill with a second drying operation being performed later by an importer, merchant or manufacturer (or at the sawmill). With certain kilns, however, it is possible to dry to a low moisture content in one operation (drying cycle). The latest research shows that drying in one cycle leads to less drying degrade and fewer stresses in the timber, and the overall costs are less.

In the case of Sweden, Finland and Norway, virtually all of the Redwood and Whitewood is dried (mostly kiln dried) before being sold by a sawmill. Most of the timber from Russia, Poland and Czechoslovakia is dried, although kiln drying is not as extensive as in the Nordic countries. Although Canadian mills are able to kiln dry, the general policy in the past has been to sell undried ('green') timber to the UK. The exception is 'Spruce-Pine-Fir' in surfaced CLS (Canadian Lumber Standards) sizes which is normally kiln dried. Discussions are taking place about further drying of Canadian timber. It is not usual for Maritime Pine from Portugal to be dried for export to the UK, or for British-grown Sitka Spruce to be fully dried at the sawmill, although it is expected that UK drying facilities, including kiln drying, will be expanded. Parana Pine from Brazil is dried for export to the UK.

MOVEMENT ON DRYING

In freshly felled timber there is 'free' water and 'bound' water. Free water is defined as water in the cells. It is called 'free' even though it is held to a large extent by capillary attraction. 'Bound' water is held in the cell walls (the cell lumina). As timber is dried, for the purpose of this book one can say that the free water is withdrawn before the bound water. The stage at which there is no free water but the bound water is retained is known as the 'fibre saturation point'. For softwoods, the fibre saturation point is generally at around 30% moisture content, although it can vary even within one species. *There is no shrinkage of timber at moisture contents down to the fibre saturation point.*

As bound water is extracted from the cell walls, the timber begins to shrink. It shrinks to a different extent radially, tangentially and longitudinally. As might be expected, there are many variations in the actual shrinkage figures quoted on the basis of various researches, and timber is much less likely to move to the same extent in subsequent changes in moisture content compared to the initial drying from green (probably because not all of the timber cross-section is likely to be affected). In the context of this book, it is usually adequate to quote the movement figures given in BS4471: 1987. They are:

- For softwoods other than those of North American origin: 'Actual sizes at moisture contents higher than 20% up to 30% shall be greater by 1% for every 5% (i.e. 5 percentage points) of moisture content and may be smaller by 1% for every 5% of moisture content below 20%.'
- For softwoods of North American origin, the movement is said to be larger at 1% for every 4 percentage points of moisture content change.

Note that the figures for shrinkage and expansion are taken as being the same for the purposes of BS4471: 1987.

In text books, the value for longitudinal shrinkage is often stated as being 'negligible'. Figures are quoted as low as 0.1% for a moisture content change of 20 percentage points (e.g. from 30% to 10%). This can well be negligible; however, differential longitudinal movement can lead to troubles with bowing of sections. In this context it is as well to be aware that shrinkage on drying along the pith and in 'juvenile wood' close to the pith, and in localized areas of 'compression wood', can be many times greater than shrinkage along 'normal' timber in the same piece. This, for example, is one reason for the desirability of excluding timber in close proximity to the pith when making door stiles or longer components which will be unrestrained in service.

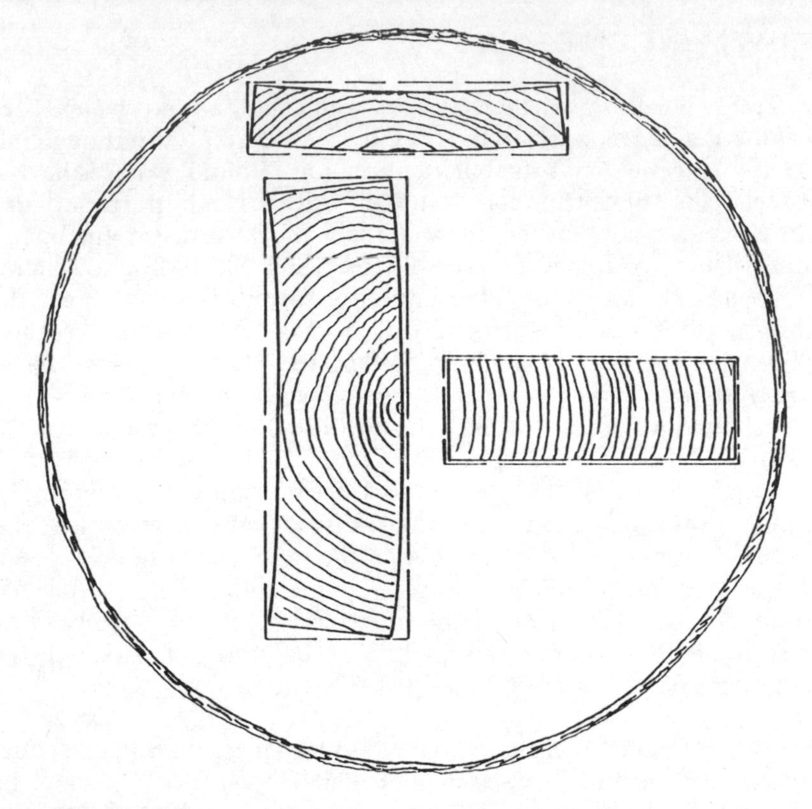

Fig. B2.1 The effect of differential radial and tangential shrinkage shown in an exaggerated way

Although the guidance on timber movement given in BS4471: 1987 is normally sufficiently accurate for most considerations of building components, Fig. B2.1 illustrates in an exaggerated way how sections sawn from different parts of a log tend to shrink due to differential radial and tangential shrinkage. One way of remembering how a section will behave is to consider that the growth rings tend to straighten-out on drying.

It is possible to restrain timber during the drying process (e.g. by clamping) and for this to reduce the initial distortion in the size as dried. If, however, a section which has an in-built tendency to distort is restrained during drying, the original stresses (even if reduced by restraining during the drying process) and those built-in by restraining may be released subsequently if the section is resawn or processed. Consequently some distortion may still occur.

METHODS OF DRYING

Although there are a few alternative forms of drying (such as dehumidification), most of the timber which is dried at a sawmill is either air dried, dried in a progressive kiln or in a chamber (compartment) kiln, whilst the majority of timber which is dried or re-dried by an importer, merchant or manufacturer is either air dried, or dried in a chamber kiln.

Three external factors affect drying and the rate of drying. These are:

- the temperature of the air;
- the relative humidity of the air;
- the speed of the air (air-flow) over the surface of the timber.

Air drying

When air drying is practiced, 'stickers' (i.e. thin pieces of dry timber) are placed between each layer of timber to be dried, so that air can circulate over all the surfaces. If thin timber boards are being dried, it may be adequate to place stickers between every second layer of boards. The stickers must be placed at sufficiently close centres so that the timber does not sag between them. Stacks are built up and placed in an open-sided shed, or left in the open with a cover placed on top to keep off the worst effects of rain, snow or direct sunshine. With air drying, however, there is little possibility to control the temperature of the air, the relative humidity of the air or the air-flow, therefore air drying is largely uncontrollable. An experienced yard supervisor may well be able to work out the best position in a storage area for certain species or sizes, for slow drying or faster drying, and can influence the air-flow to a limited extent by selection of the sticker thickness and the disposition of the stacks, but still the rate of drying will be influenced largely by factors outside the control of yard staff.

Air drying ceases when the temperature is low, which is one of the main reasons why sawmillers in countries which have a cold winter install kilns if they wish to operate and sell dried timber throughout the year.

If a timber species which is prone to sap stain (see later notes on sap stain) is to be air dried, it is desirable for it to be treated with an anti-stain solution soon after being sawn from the log because, with air drying, it could be several weeks or even months before the moisture content is brought below the level (about 20/22%) above which sap stain can develop during this period.

The moisture content of air dried timber is dependent on the relative humidity and temperature of the air. In the UK and Northern areas, air drying of thicknesses of 38 mm and over is unlikely to yield moisture contents of less than 18/20% in a reasonable time (although surface readings may be lower).

For softwoods, Table B2.3 shows the relationship between relative humidity, temperature, and the equilibrium moisture contents likely to be met in practice.

Progressive kilns

Progressive kilns are the type of kiln most common in the larger Nordic and European sawmills. The largest types have separate channels for each thickness of timber. Each channel can be about 25 m long, 4 m high or higher, and 6–7 m wide.

Sawn softwood is 'sticked' by automatic machinery before stacks are presented at the entrance to a channel. Timber stacks enter at one end of a channel (with timber lengths parallel to the width of the channel), and move along on trolleys as new stacks are fed in and as dried stacks are taken out. With most progressive kilns, air of controlled temperature and humidity is fed in only at the entry position. The operation and control of the kilns is by automatic process, with checks being made on the timber as it progresses along each channel.

With kiln drying it is necessary to create high humidity at the in-feed position so that the surfaces of the timber are not dried before moisture is drawn out of the centre of each piece. The heat for kilning can be produced by oil, but some sawmills try to be self-sufficient in fuel by burning bark, chippings, sawdust or even branches.

The most efficient way to use a progressive kiln is to keep each channel continuously full of one species of timber of one thickness or similar thicknesses. This method of operation allows the correct kilning cycle for one particular thickness to be used, and results in the highest possible quality in the dried timber. Unfortunately this cannot always be arranged when a mill is trying to give service, particularly if it is repeatedly requested to provide many small quantities of different sizes.

Softwood thicknesses of 63 or 75 mm take in the order of 150/200 hours depending on the kiln design to be dried to an 'average' of about 20% moisture content, whilst thicknesses of 19–25 mm take about 60/120 hours to finish at a lower average moisture content. 'Average' in this context is a statistical average for the number of pieces in a stack.

The spread of results in the majority of pieces in one batch of timber as removed from a progressive kiln is likely to be in the order of plus or minus 3 percentage points or, more likely, plus 4 minus, 2 percentage points until the stacks have 'conditioned' (i.e. adjusted to a more uniform moisture content as affected by the relative humidity and temperature of the surrounding air). If, however, one measures the moisture content of many batches or stacks as removed from a kiln, it is possible to encounter different average values. The values given in Table B2.1 are based on extensive research carried out in Sweden.

This general level of drying is sometimes referred to as 'shipping dry'. As can be seen, the primary object of drying the thicker pieces to this level is to achieve the same general level as is obtained by air drying. By the use of kilns, however, the timber can be dried

Table B2.1 Average values and spread of results of kiln dried timber

Average values and spread of results for 95% of all pieces		
Timber thickness (mm)	**Average (%)**	**Spread of results (%)**
19–25	15	10.6–19.4
38–50	18	14.0–22.0
63–75	20	16.4–23.6

Average values and spread of results for 95% of pieces in the driest and wettest batches as removed from a progressive kiln		
Timber thickness (mm)	**Average (%)**	**Spread of results (%)**
Driest batch		
19–25	10	7.2–12.8
38–50	15	12.6–17.4
63–75	16	13.6–18.4
Wettest batch		
19–25	17	10.0–24.0
38–50	20	14.8–25.2
63–75	22	18.0–26.0

throughout the year, whereas air drying stops in cold weather.

To dry timber, it is necessary for there to be a 'moisture gradient' between the centre of the cross-section and the surfaces (drier on the surfaces). One must expect a certain moisture gradient to remain in timber when it is first removed from a kiln, even though the kiln cycle is designed to keep the gradient to a minimum by including a conditioning period at the end of the schedule. When kiln-dried timber is stored, the moisture content will gradually equalize, depending on the temperature and relative humidity of the surrounding conditions.

After drying, the timber is de-sticked automatically and is then graded.

The maximum temperature in the kilning cycle of a typical Nordic progressive kiln has normally been about 50/55% or lower, however, experiments with slightly higher temperatures are giving encouraging results in yielding quality timber with fewer fissures, providing that other aspects of kilning are carefully controlled.

Although progressive kilns can be used to dry softwoods to moisture contents of less than 'shipping dry', 'chamber kilns' are regarded as giving a better means of obtaining low moisture contents.

Chamber kilns (or compartment kilns)

In a chamber kiln, the timber stacks are not moved during the drying cycle. As with a progressive kiln, the timber is 'sticked' to allow maximum air circulation during drying. Chamber kilns can be quite small (e.g. 2.4 m wide, 2.4 m high, and 12 m long) although, at sawmills or large joinery factories, kilns having chambers in the order of 8 m wide or wider, 7 m high and 9 m deep are quite normal.

Chamber kilns can be used to dry softwoods to a moisture content as low as 8%, and usually achieve a spread of results of about plus or minus 2 percentage points at the 15% average level, and even closer at lower levels. It is worth noting that, although it is more difficult to kiln dry to low moisture contents, the spread of results is less than at the higher moisture content levels.

For years, discussions took place on whether or not it is better to dry timber to a low moisture content by drying in two stages with a conditioning period in between, or by drying in one operation. Either way is possible, but obviously the overall cost is less if sticking, de-sticking, handling and grading are carried out only once. It is fortuitous, therefore, that the most recent research shows quite conclusively that there is less overall drying degrade, and more uniformity of moisture content, if kiln drying can be arranged to be carried out in one operation commencing with green timber. Sometimes, however, a sawmill chooses to kiln dry in two stages so that a selection process can be carried out after a standard stock size has been dried to the normal level, before drying the selected pieces (perhaps of higher quality than the rest) to a lower moisture content.

Indicative times for drying 25 mm and 75 mm thicknesses to low moisture contents are given below for cases where the quality of the dried timber is important. The actual times will vary with kiln design and condition.

To dry 25 mm thick green timber to 8% moisture content in one operation (cycle) is likely to take about 7–8 days, whereas to dry in two operations from green to 18% and then from 18% to 8% is likely to take about 3–5 days for the first cycle and about 6 days for the second. The beginning of the second cycle must include a stage of increasing the relative humidity and temperature (steaming/conditioning) which has a significant influence on the time taken.

Currently it is not thought commercially practical to dry 75 mm thick green timber to 8%. To dry 75 mm to 10–12% in one cycle is likely to take about 14 days, whereas to dry in two operations from green to 18% and then from 18% to 10–12% is about 6–8 days for the first cycle and about 12 days for the second. As stated above, however, different kiln designs require different drying cycles.

High-temperature kilning

Where timber is not to be resawn before use (e.g. for structural purposes), some mills (e.g. in Canada and New Zealand) use high-temperature kilning. A typical cycle requires temperatures of up to 105°C and will kiln 38 mm thick softwood in under 50 hours. Temperatures as high as 115°C have been used and reduce the kilning time to 25/27 hours. Almost inevitably, immediately after drying, the timber has a large moisture gradient and may be 'case-hardened' (see later notes on case-hardening in this chapter).

Sap stain and anti-stain

Some timbers (e.g. European Redwood) are prone to sap stain (sometimes called 'blue stain' or 'yard blue') in the sapwood if it remains wet. One stage at which sap stain can form is during the time soon after sawing timber from a log. This may occur in some species at certain warm times of the year if the time between sawing and the commencement of drying is more than about 5 days. Sap stain can also occur if sawn timber is subsequently allowed to remain at a moisture content of over 20/22% for a period of a few weeks.

Sap stain is not a structural defect and is permitted by BS4978: 1988 for stress graded timber, but can be considered to be a visual defect which reduces the value of the timber. BS1186: Part 1: 1986 and the latest draft of BS1186: Part 3 limit the extent to which 'discoloured sapwood' is permitted in joinery and wood trim.

To avoid sap stain, some mills dip the timber in an anti-stain solution before the timber is dried. In the past, the most common of these solutions were based on pentachlorophenol (p.c.p.), but some countries have a full or partial ban on the use of p.c.p. because of its possible adverse effects on the environment at the place of application, and the difficulty in ensuring that waste is disposed of safely if the timber is subsequently processed. Some safer anti-stains have been and are being developed, but many mills do not use anti-stain at all, preferring to avoid stain whilst the timber is within their control by moisture control and protection. As an alternative, some mills use an anti-stain only in warmer weather when sap stain is more likely to commence if the moisture content is sufficiently high.

Sap stain will not occur after timber has been dried correctly unless the moisture content is allowed to increase again and to exceed about 20/22% for a few weeks. If sap stain has commenced, it will not progress further once the moisture content falls below about 20/22%, but the existing stain will remain.

An anti-stain treatment is intended to be effective only for a period of a few months, not for years. The intention is to prevent sap stain occurring during the initial storage at a mill, and during transportation from a mill to the initial buyer, not for it to act as a long-term preservative.

MOISTURE CONTENTS IN SERVICE, AND MAXIMUM MOISTURE CONTENTS

There are variations in figures quoted for the moisture contents

Table B2.2 Typical moisture contents encountered in service

	Average values (%)
Joinery inside buildings with continuous heating giving a room temperature between 20 and 24°C	8–12
Joinery inside buildings with continuous heating giving a room temperature between 12 and 19°C	10–14
Joinery inside buildings with intermittent heating	13–17
External joinery such as windows to heated or unheated buildings	13–19
Structural components in a continuously heated building	14
Structural components in a building, covered and generally heated	16
Structural components external, fully exposed	18% or more (sometimes considerably more)

encountered in service, but those given in Table B2.2 can be regarded as indicative and as typical guidance for softwoods. They are based on BS1186: Part 1: 1986 on joinery and the structural design code BS5268: Part 2: 1988.

In addition, BS1186: Part 1: 1986 states that no individual reading of moisture content of softwood should exceed 23% for exterior joinery and 20% for interior joinery at the time of handover to the first purchaser, and BS5268: Part 2: 1988 states that, at the time of erection, no moisture content reading should exceed:

- 24% for timber 'covered and generally unheated' in a building;
- 21% for timber 'covered and generally heated' in a building;
- 19% for timber 'internal in a continuously heated building'.

As drafted and released, in 1989, for publication by BSI, BS1186: Part 3 on wood trim gives the same moisture content figures as in Table B2.2 for joinery. It gives them, however, as a range between permitted limits not as average values, and does not permit higher values for some readings as is permitted in BS1186: Part 1: 1986. For example, the limits for exterior wood trim are 13% minimum and 19% maximum.

Measuring moisture content

If one measures moisture contents in laboratory conditions, it is possible to adopt sophisticated methods which take account of most of the volatile extractives which evaporate during drying, but normally two methods are considered to be sufficiently accurate to measure the moisture content of joinery, wood trim, carpentry or structural components. These methods are the oven-drying method and the use of a moisture meter.

Moisture content is normally expressed as a percentage of the weight of moisture in the timber compared to the weight of timber dried in an oven to constant weight. (Note, *not* the weight of moisture in the timber compared to the total weight of the wet timber.)

$$\text{Moisture content} = \frac{\text{weight of moisture in the timber}}{\text{weight of oven-dried timber}} \times 100\%$$

$$= \frac{\text{wet weight} - \text{oven-dried weight}}{\text{oven-dried weight}} \times 100\%$$

Oven-drying method

With the oven-drying method of measuring moisture content, it is necessary to cut a sample or samples from the timber being measured, therefore it is a 'destructive' method, particularly because it is necessary to cut samples at least 200/250 mm from the end of a piece. The oven-drying method is often said to be more accurate than the use of a moisture meter. However, it is as well to realize that the answer is an average and only for the particular sample (which is one reason why samples must be small), is accurate only for the time at which the sample is taken even though the result takes time to process, and can be affected by resin and extractives in the timber which are removed in the oven.

Measuring by a moisture meter

Measuring timber by a moisture meter is a way of obtaining an instantaneous non-destructive reading. It must be realized that each reading is local to the probes. If short probes are used, a surface reading is obtained. If longer insulated probes are used, each reading is of the moisture content at the tips of the probes, not an average over the length of the probes. Therefore, one use of a moisture meter is to plot the moisture gradient through a piece of timber. Indeed, a moisture meter can be used to trace the source of moisture ingress to a component in a completed building. Normally, to satisfy a product standard, it will be necessary to take several readings and to ensure that the average and the upper readings (and perhaps the lowest reading) fall within stated limits.

Most meters have several graduated scales and it is important to use the one intended for the species being checked.

Although the method is non-destructive, it should be realized that the holes made by deep probes can be noticeable on an exposed face of joinery if the holes are not to be filled and covered by an opaque paint. Therefore, for joinery checked before installation in a building, it is preferable to insert the probes in a surface which will be concealed in the final work.

Standards call for moisture meters to be calibrated regularly by the meter manufacturer. In addition, it is worth noting that the UK Timber Research and Development Association developed a small resistance device which can be used to check the meter on each occasion before readings are taken.

The principle on which a moisture meter works is to measure electrical resistance between the probes, therefore any additive between the probes will affect the resistance and hence the reading. CCA preservatives or flame retardants in particular, affect the resistance in the timber and, when they are present, the meter will not give a true reading of the moisture content of the timber. The manufacturer of the meter may be able to provide a correction factor. Organic solvent preservatives are considered not to affect readings to any significant extent, with the possible exception of those containing copper naphthanate.

Measuring relative humidity and temperature

Table B2.3 shows the relationship between relative humidity, temperature, and equilibrium moisture content of softwoods. Fairly recently, meters have been developed which give an instantaneous reading of relative humidity as well as temperature. These are extremely useful, for example, if one is trying to determine if the conditions within a factory or store are suitable for a timber item or are too dry or damp. From a study of the values given in Table B2.3, it is possible to see how important relative humidity is in determining the moisture content of timber and in controlling or determining the 'equilibrium moisture

Table B2.3 Equilibrium moisture contents of softwoods

Relative humidity (%)	Temperature (°C)							
	16	18	20	22	24	26	28	30
85	18	18	18	18	18	18	18	18
80	17	16	16	16	16	16	16	16
75	15	15	15	14	14	14	14	14
70	13	13	13	13	13	13	13	13
65	12	12	12	12	12	12	12	11
60	11	11	11	11	11	11	11	10
55	10	10	10	10	10	10	10	10
50	9	9	9	9	9	9	9	9
45	9	8	8	8	8	8	8	8

content'. Normally it has more effect on the equilibrium moisture content than temperature.

When measuring moisture content in a factory, it is important to realize that the equilibrium value is not necessarily the actual moisture content of timber which has recently been moved into the particular environment. Timber at a higher or lower moisture content takes time to adjust to the prevailing relative humidity and temperature conditions.

POSSIBLE DRYING DEGRADE

Moisture gradient

When green timber begins to dry, a 'moisture gradient' is set up between the centre and surface of the timber. A gradient is necessary if the timber is to dry but, to minimize degrade, the gradient should not be too severe or surface splitting and/or 'case-hardening' (see later notes) can occur. Also, it is easier for moisture to be extracted from end grain than faces and, if this is not controlled, end splitting can occur. (When selling 'green' timber, some mills paint the ends of timber to give a partial seal to reduce the extent of end splitting due to uncontrolled rapid drying out.)

With kiln drying, it is possible to control the rate at which moisture is drawn from the surface of the timber by enveloping the timber at the beginning of the drying cycle with air having a high relative humidity. By controlling heat and humidity and using the right balance throughout a drying cycle, timber can be dried with minimum checks or splits.

Resin exudation

When drying resinous timbers, if the maximum temperature is limited to around 50°C, resin exudation during the drying process is minimized, but not eliminated.

Areas of high-porosity or over-porosity

Localized areas of high-porosity may be encountered on the surface of dried timber. If a dark stain is used, it may highlight such areas by the amount of stain and hence colour pigment which is absorbed being much more than that absorbed by the surrounding timber. Areas of high-porosity are more likely to be encountered with certain species (e.g. European Redwood) if the logs from which the timber was cut were stored de-barked in water for a considerable time during warmer times of the year. Such high-porosity or over-porosity is caused by biological action which, however, is not considered to weaken the structure of the timber. Companies are experimenting with additives to organic solvent preservatives with the intention of sealing end grain and areas of high-porosity. Also, Nordic mills are aiming to eliminate water storage of logs, particularly Redwood logs. When necessary to limit degrade due to premature drying-out, land stored logs arc sprayed with water. With some species, even the period of spraying must be limited in warm weather to a few weeks to prevent the development of surface areas of high-porosity.

Case-hardening

The term 'moisture gradient' should not be confused with 'case-hardening'. If drying is not controlled correctly in the early stages, and surface moisture is extracted too quickly, case-hardening may result. This can occur when a surface dries out before the centre of the timber, to such an extent that the surface layer cannot shrink because the centre is restraining it. At this stage, the surface is in tension and the timber at the interface with the un-dried part of the centre is in compression. If the strains which are caused at the surface exceed the elastic limit, some permanent set in the surface layer occurs. Later, when the centre has dried out fully and the centre has shrunk more than the permanently set surface, the stress at the surface will have reversed to compression, and the timber at the interface with the centre will be in tension. It is normal during air or kiln drying for stresses to be set up temporarily but not of such a magnitude as to cause permanent set or case-hardening.

Case-hardening is rather an unfortunate term because the surface is not made harder and there is no 'case' effect, therefore the permeability of the timber stays the same. Little is written in timber technology books about case-hardening in softwoods. This is probably because it is not common in softwoods, and because even quite severe case-hardening does not normally lead to serious trouble with softwoods unless the cross-section is deep sawn (i.e. sawn to divide the thickness into two or more pieces), in which case 'cupping' may occur. (With heavily case-hardened hardwood, a form of cell collapse known as 'honeycombing' may occur, but extensive reading has encountered no mention of anything similar occurring with softwood.)

A parcel of timber which is case-hardened can normally be corrected by conditioning in a kiln at high humidity and temperature, providing that the conditioning is carried out before the timber is resawn or moulded.

If it is decided to test for case-hardening, the usual test method is to

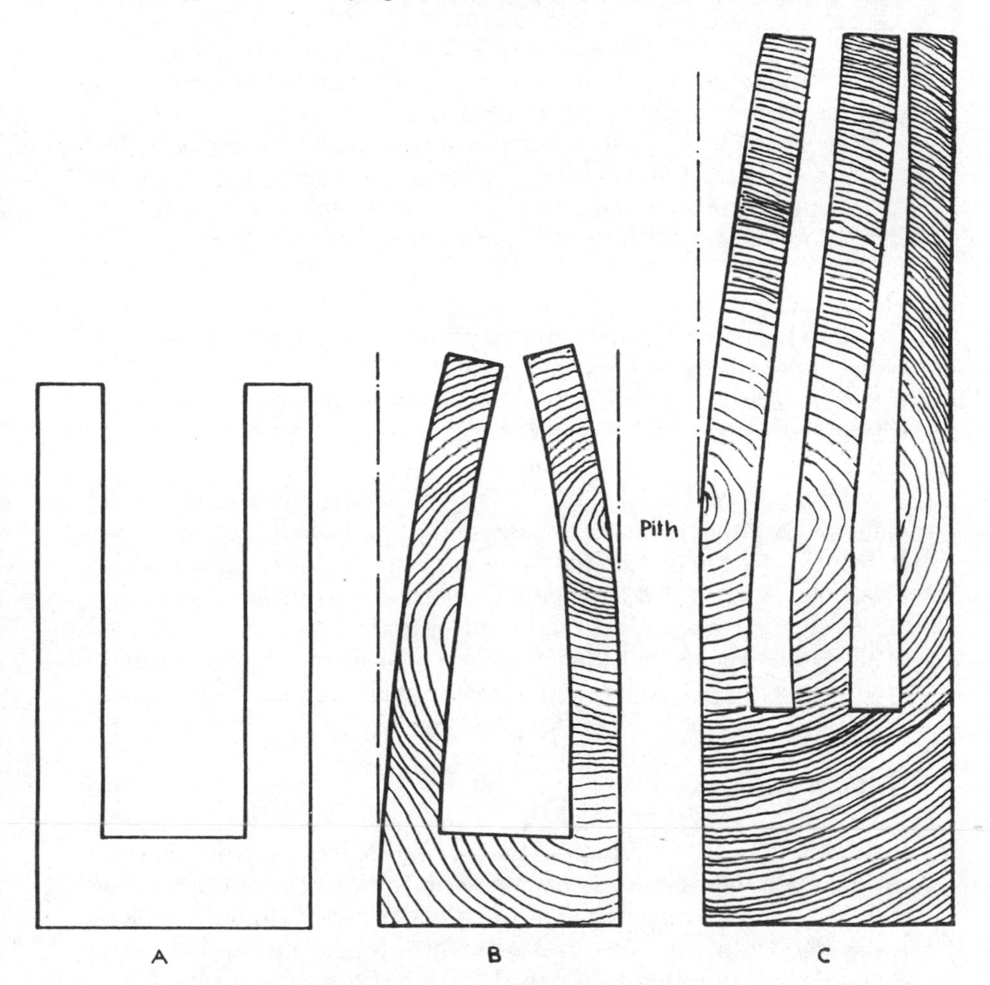

Fig. B2.2 Prong test pieces to check if there is case-hardening

cut a cross-sectional slice some 12 mm thick at least 200 mm from the end of a piece. This is then cut to the general 'prong' shape shown in A of Fig. B2.2. The operative cutting the shape should be instructed to note if the timber 'grips' the saw during cutting out the centre. If this happens, it is a sign of heavily case-hardened timber. If no gripping occurs it is likely that there is no significant case-hardening.

If the prong test is carried out on a sample cut from timber in the early stages of drying, the prongs will probably first curve outwards even if the drying is being carried out correctly. However, the test piece must then be dried to uniform moisture content. If *both* prongs finally curve inwards as shown in B of Fig. B2.2 this is a sign of case-hardening. If the prongs are straight, or more or less straight, there is no case-hardening.

If the prong test is carried out on a sample cut from timber at a later stage of drying, the prongs may well curve inwards straight away. If, after drying to constant moisture content, *both* prongs are still curved inwards, this is a sign of case-hardening.

If the prong test piece is cut from timber which includes or is close to the pith, in analysing the test, care must be taken to note the shape of both prongs in relation to the position of the pith. If only one prong curves (or only one outer prong as shown in C of Fig. B2.2) and it is on the same side as the pith, with the other prong (or the other outer prong) being straight, it is almost certain that there is no case-hardening and that the curve in one prong is due purely to differential radial/tangential shrinkage. The sketches B and C in Fig. B2.2 are traced from two actual prong tests. The growth rings shown are as they were in the actual samples and show the position of the pith. In Sample B it can be seen that both prongs are curved inwards, indicating case-hardening. It is worth noting, however, that in the prong which contains the pith, the curve increases at the point where the pith occurs. In Sample C it can be seen that only one outer prong is curved. This is the prong which includes the pith. The other outer prong is as straight as one could expect and is at right-angles to the edge. The straightness of this prong shows that there is no case-hardening. The curve in the other outer prong is due to differential radial/tangential shrinkage.

Growth stresses

If stresses have been built into a tree during its growth, particularly its early growth, they are likely to remain in some pieces of timber sawn from the tree and can result in distortion in the sawn timber as it dries. If it is known that the species is one which is prone to distort during drying, one way to limit distortion is to restrain the sections during drying. It should be realized, however, that the stresses (even if reduced by the restraints used during the drying process) are still locked-in and, if the sections are resawn, the stresses may be released and may still lead to some distortion. If the sections are not resawn or heavily moulded, and are restrained in service, they are unlikely to distort with time. Current research into kiln drying cycles is aimed at reducing built-in stresses and their effect.

PROTECTING DRIED TIMBER

Every code of practice which deals with the use of dried timber emphasizes the need to protect it during storage, transport and erection. This is even more important if timber dried to low moisture contents is used. Basically sufficient level supports, cover from weather, and ventilation are required. It is as well to realize, however, that once softwood has been dried correctly, even a heavy rain storm on unprotected timber is likely to affect only the outer few millimetres of surface (although end grain penetration may be more). In this context, it is relevant to point out that European Whitewood is particularly resistant to the ingress of moisture (although behaves normally to the passage of moisture vapour).

It takes a purpose-built kiln several days to remove moisture from timber, therefore it is unlikely that a rain storm of a few hours will force moisture back to any great extent. This is one case where a moisture meter with insulated probes can be used to show just how far moisture has penetrated. After a rain shower, a surface reading of moisture content will give a misleading impression of the moisture content inside the timber.

INSTALLING TIMBER

Items of joinery, wood trim etc. should be installed at moisture contents close to the service moisture content (see Table B2.2), and not in excess of the 'maximum moisture contents' quoted in the text below Table B2.2. Similarly, structural timber should be installed at moisture contents close to those given in Table B2.2, and not in excess of the 'maximum moisture contents' quoted below Table B2.2.

If, for whatever reason, a member which will be subject to bending stresses is found during construction to have a moisture content significantly higher than 18/20% m.c., load should not be applied until the moisture content has been reduced to 18% or lower. If the member is a beam, and load has been applied before the high moisture content is discovered, consideration should be given to propping the beam until the moisture content has reduced, or else 'creep deflection' is likely to take place as the timber dries out.

STRESS GRADING B3

INTRODUCTION

When timber is to be used structurally it must be graded for strength, even when the structural use is a simple one such as a floor joist. To satisfy the relevant British Standard BS4978: 1988 on softwood grades for structural use, and most other stress grading standards, the intention is that each piece of stress graded timber, or the component which contains the stress graded timber, should be marked to indicate the strength. Certainly it must be possible for building control officers and similar control staff to be able to check from the marking if structural timber has been stress graded, the level to which it has been graded, the species, and also to identify the company or grader responsible for the grading. Where stress graded timber is to be left exposed and there is a specific requirement for marking to be excluded, BS4978: 1988 permits an alternative to marking which is a form of written certification.

Until around 1970, the only known commercial method of stress grading was by visual inspection. Therefore certain limits of knot size, fissures etc. had to be detailed for each stress grade. Since then, however, stress grading machines have been developed and have been proved in practice to be able to stress grade sawn or processed timber in either the dried or undried ('green') condition.

Until an investigation of the strength of softwoods, begun in 1968 as a collaboration between the laboratories of the UK, Sweden, Finland and Canada, most of the research on strength had been carried out on small clear specimens of undried timber to establish a 'basic' stress. Coefficients were then applied for green and dry timber to take account of the anticipated strength reducing effects believed at that time to be due to knots, fissures etc. More recent laboratory work, helped tremendously by the availability and speed with which strength data can be accumulated by using stress grading machines, has been carried out on commercial sizes of timber containing the normal characteristics and defects encountered in practice. This has shown the previous assumptions on strength to have been generally too conservative as far as bending and compression is concerned, although the work showed that timber graded in certain ways is generally weaker in tension than had been believed. With regard to knots, previous strength reducing coefficients in the UK design code took account of the way in which knots affected the surface of a piece. From Fig. B3.1 it can be seen that several types of knot having the same size on a surface can affect the cross-section and strength in different ways. These differences cannot be taken into account only by examining the way in which a knot affects one surface. Now it has been established, however, that an experienced visual stress grader can estimate the way in which a type of knot (or knots) affects the cross-section of a piece with sufficient accuracy for this knowledge to be built into the more recent 'knot area ratio' method grading rules. The way in which knots affect a cross-section has more relevance to strength than the way in which they affect a surface. In particular, reference to Fig. B3.1 shows that conical knots do not affect the strength to the same extent as was assumed in the past in the UK, when UK stress grading rules and design stresses were based on the assumption that knots, such as those shown in Fig. B3.1, had parallel sides.

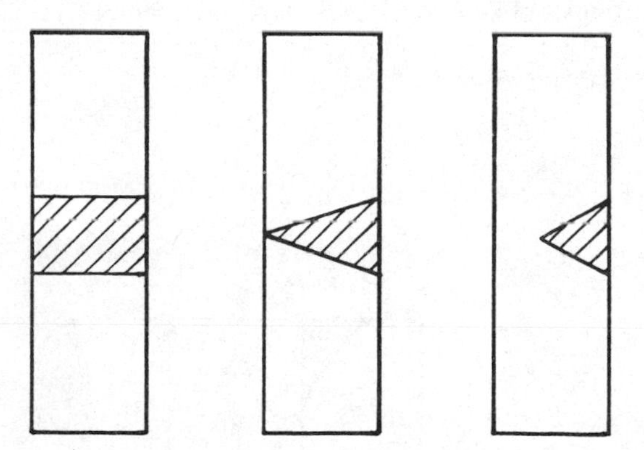

Fig. B3.1 Different ways in which knots with the same surface area can affect cross-sectional area

VISUAL STRESS GRADING

With the BS4978: 1988 method of visual stress grading, the limiting sizes of knots are based on what is called a 'knot area ratio' (KAR). This is the ratio of the extent to which the projected area of a knot (or knots) affects the area of the cross-section of a piece rather than the surface of a piece. Obviously the grader has to be given ratios which are

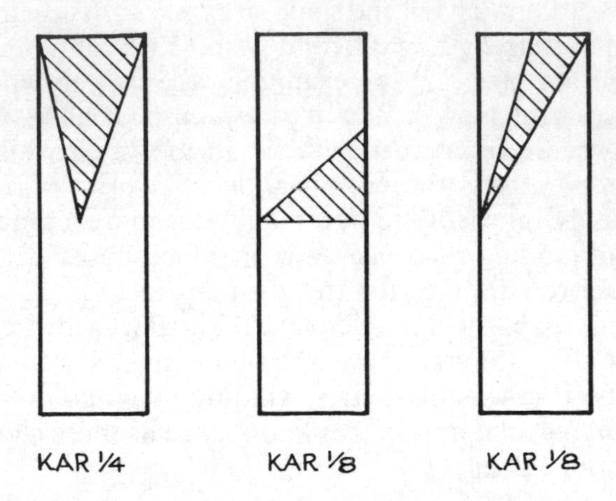

Fig. B3.2 Examples of knot area ratios easy to remember

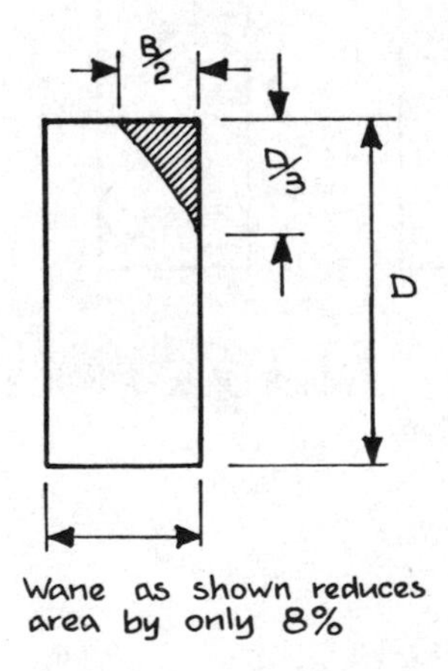

Fig. B3.3 The limited effect of wane on cross-sectional area.

easy to remember and estimate (one half, one third etc.). Figure B3.2 illustrates knot area ratios which are easy to estimate.

With visual stress grading, there are other limits which must be observed for fissures, wane, slope of grain, rate of growth, abnormal defects, insect attack and decay. Although sap stain can look unsightly if there is a great deal of it, it will not spread when the moisture content of the timber is below about 20/22% and does not reduce the strength of the timber, consequently it is permitted in timber stress graded to BS4978: 1988. Wane is limited in grading rules but is mainly significant at points of bearing. As can be seen by reference to Fig. B3.3, even very noticeable wane leads to only a small percentage reduction in the area of a piece. Therefore limits are decided largely for cosmetic rather than strength reducing reasons (other than at bearing points and connections).

BS4978: 1988 describes two visual stress grades, 'General Structural' and 'Special Structural' (Special Structural being the stronger) which are abbreviated to GS and SS particularly when marking the timber. (When machine stress graded the marks are MGS and MSS.) As well as adopting the knot area ratio method for visual stress grading, BS4978: 1988 differentiates between knots near an edge and the centre of a rectangular section (but not a square section) by introducing the concept of a 'margin'. Knots near an edge have more effect on the strength of a piece than knots near the centre. Figure B3.4 illustrates margins which, in the case of BS4978: 1988, are the outer quarters of area. Because a grader will not know how a piece will be fixed, BS4978: 1988 does not differentiate between the effect of knots in the upper or lower margin but, for solid timber joists, the assumption is that sections will be fixed to take load on the strong axis.

A 'margin condition' is said to exist when more than half of the area of one margin is occupied by the projected area of a knot or knots. In

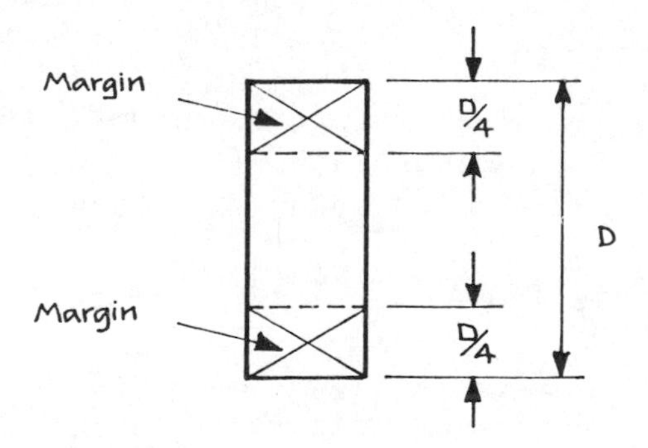

Fig. B3.4 Margins as defined in BS4978: 1988

such a case, the total permitted knot area ratio is smaller than that for the cases where a margin condition does not exist. For example, in the case of a rectangular piece of General Structural graded to BS4978: 1988, where a margin condition exists, the total knot area ratio is limited to one third whereas, where a margin condition does not exist, the total knot area ratio is limited to one half. For Special Structural, the values are one fifth and one third respectively. The knot area ratio limit for a square piece is the tighter of the two limits (e.g. for GS it is one third).

One disadvantage of visual stress grading is that the method does not take account of density. Density has a significant effect on strength, therefore, because visual stress graders cannot measure density accurately, permissible stresses for visual grades are quoted on the safe side from a knowledge of the density spread within each species. Stress grading machines take account of density and, consequently, when machine stress grading to the same level as a corresponding visual grade, normally there will be a higher yield from the same consignment of timber (unless of course the timber being machine stress graded has unusually low density for the species).

Before a designer can allocate grade stresses to a piece of visually

Table B3.1 Comparison of BS5268: Part 2: 1988 stresses for BS4978: 1988 stress grades

	Grade bending stress dry (N/mm²)[1]				Mean E value dry (N/mm²)			
Species	GS MGS	M50	SS MSS	M75	GS MGS	M50	SS MSS	M75
European Whitewood or Redwood from Sweden, Finland, Russia, Poland, Czech., Norway	5.3	6.6	7.5	10.0	9000	9000	10500	11000
Canadian Spruce-Pine-Fir	5.3	6.2	7.5	9.7	8500	9000	10000	10500
Canadian Hem-Fir	5.3	6.6	7.5	10.0	9000	10500	11000	12000
Canadian Douglas Fir-Larch	5.3	–	7.5	–	10000	–	11000	–
British-grown Sitka Spruce and British-grown European Spruce	4.1	4.5	5.7	6.6	6500	7500	8000	9000

[1] Not modified by the width (depth) factor of BS5268: Part 2.
– No figures available.

stress graded timber, the species or species mix must be known because different species have different strengths. For the UK, stresses are tabulated in design code BS5268: Part 2: 1988. Dry grade bending stresses and mean E values for the major softwoods likely to be available in timber stress graded to BS4978: 1988 are given in Table B3.1 to indicate the comparison between species. It is not normal for Parana Pine or Canadian Western Red Cedar to be available stress graded, therefore stresses are not quoted in Table B3.1 for these species (although they are tabulated in BS5268: Part 2: 1988).

Stress grading to BS4978: 1988 is a 'full length grading', with the permissible stresses based on the weakest cross-section of a piece. Larger defects are not permitted at ends as is the case with some grading rules. Therefore, if a piece is cut into two or more separate lengths, the grade of each length is not reduced. If, however, the piece is resawn in cross-section, the stress grading is invalidated and each piece must be regraded if to be used structurally.

Visually stress graded laminates for glulam

BS4978: 1988 describes three stress grades: LA, LB and LC for laminates for glulam sections (although the top grade LA is rarely used because of the difficulty of providing full strength end joints commercially for this high grade). With visual stress grading, the measurement of knots is by the KAR method, but no margin or margin condition is considered.

Availability of GS and SS visually stress graded

GS and/or SS in European Whitewood or European Redwood which has been imported from Sweden, Finland, Russia, Norway, Poland or Czechoslovakia is readily available. It is more usual to stress grade Whitewood than Redwood although Whitewood and Redwood are mixed in some parcels. The stress grading can be carried out in the country of origin or in the UK. The same rules apply in either case. Several Swedish and Finnish mills and a few Polish, Norwegian, Danish and Canadian mills have been authorized to grade to GS or SS and are stress grading at source. Until 1988, BS4978 did not insist on visual stress grading of GS or SS being carried out by authorized graders, but obviously it is better for trained graders to be used and, from 1988, both visual and machine stress grading is required to be carried out within a recognized quality assurance scheme. Such schemes recognized by the UK design code committee (not only for grading to BS4978) are referenced in an Appendix to BS5268: Part 2: 1988. (During 1989, 'The UK Timber Grading Committee' was established to take over from the UK design code committee the responsibility of approving grading agencies.) The 'TRADA Quality Assurance Services' company licences graders to grade GS and SS in the UK and, currently, in Norway and Denmark. The relevant authorities in Sweden, Finland and Poland, and Canadian grading agencies arrange the training and approval of graders to grade GS and SS in their country.

Canadian visual stress grades

Several Canadian stress grades visually graded to the NLGA (National Lumber Grades Authority) rules are tabulated in BS5268: Part 2: 1988 with the permissible stresses allocated in the UK. Grade descriptions vary with the size of the timber.

For sizes of 38 × 114 mm and larger (see Tables B1.2, C5.1 and C5.3) the relevant stress grades are called 'Structural Joists and Planks'. Within this description the various stress grades are:

'Select Structural Joists and Planks'
'No. 1 Structural Joists and Planks'
'No. 2 Structural Joists and Planks'
'No. 3 Structural Joists and Planks'.

These grades are applicable both for sawn timber and surfaced CLS sizes of 38 × 114 mm and wider.

In the UK it used to be normal to buy No. 1 and No. 2 Structural Joists and Planks mixed in a parcel. However, the 1988 edition of BS5268: Part 2 gives the same stresses for No. 1 and No. 2. As a consequence, No. 1 is not normally sold in the UK. For CLS, Select Structural has become more generally available, is marked separately, but is sold with No. 2 in a mixed parcel. Select Structural is also more readily available in sawn sizes.

For sizes of 38 × 89 mm, 38 × 63 mm or 38 × 38 mm (see Table C5.2), the relevant stress grades are:

'Select Structural Light Framing'
'No. 1 Structural Light Framing'
'No. 2 Structural Light Framing'
'No. 3 Structural Light Framing'

'Construction Light Framing'
'Standard Light Framing'
'Utility Light Framing'

'Stud Grade'.

The stress grades most commonly available in the UK in the size 38

Table B3.2 Comparison of BS5268: Part 2: 1988 stresses for Canadian stress grades

	Grade bending stress dry (N/mm²)			Mean E value dry (N/mm²)		
	Spruce-Pine-Fir	Hem-Fir	Douglas Fir-Larch	Spruce-Pine-Fir	Hem-Fir	Douglas Fir-Larch
'Structural Joists and Planks' (Sizes of 38 × 114 mm and over)						
'Select'	**8.0**	**8.0**	8.0	**10500**	**12000**	12500
'No. 1'	5.6	5.6	5.6	9000	10500	11000
'No. 2'	**5.6**	**5.6**	5.6	**9000**	**10500**	11000
'No. 3'	4.1	4.1	4.1	9000	10500	10500
'Structural Light Framing' (Stresses for 38 × 89 mm only)						
'Select'	**9.1**[1]	**9.1**[1]	9.8[1]	**10500**	**12000**	12500
'No. 1'	6.4[1]	6.4[1]	6.4[1]	9000	10500	11000
'No. 2'	**6.4**[1]	**6.4**[1]	6.4[1]	**9000**	**10500**	11000
'Light Framing' (Stresses for 38 × 89 mm only)						
'Construction'	**5.4**[1]	**5.4**[1]	5.4[1]	**9000**	**10500**	11000
'Standard'	**4.0**[1]	**4.0**[1]	4.0[1]	**9000**	**10500**	10500
'Utility'[2]	3.2[1]	3.2[1]	3.2[1]	9000	10500	10500
'Stud Grade'[2] (Stresses for 38 × 89 mm only)	4.7[1]	4.7[1]	4.7[1]	9000	10500	10500

The values shown as bold indicate the grades most likely to be available.
[1] Stresses already increased (by 14.3%) for the width (depth) factor of BS5268: Part 2.
[2] Not usually imported to the UK.

× 89 mm used to be a mix of 'Construction/Standard Light Framing', but permissible stresses for these grades were reduced in BS5268: Part 2: 1988 and there is a swing towards the use of the 'No. 2 Structural Light Framing'. 'No. 3 Light Framing', 'Utility Light Framing' and 'Stud Grade' are not normally available in the UK. For sizes smaller than 38 × 89 mm the same stress grade descriptions apply but there are coefficients which vary the permissible stresses.

For comparison purposes, Table B3.2 gives the dry grade bending stress and mean E value of various Canadian visually graded stress grade/species combinations, with the combinations likely to be more readily available shown in bold type.

MACHINE STRESS GRADING

It has been established that there is a reasonably accurate relationship between the stiffness of timber and its strength in bending. It has also been established that it is possible to measure the stiffness on the weak axis of a piece of timber and relate this to its strength on the strong axis. This is because density has a large influence on both stiffness and strength. Once the existence of the relationship between stiffness and strength was established, it became possible to develop non-destructive machines to measure the stiffness of timber as it is passed through a machine and, from this, to allocate a strength rating to each piece of timber. Machines were developed and have been used commercially since the early 1970s by manufacturers and importers in the UK, at sawmills in Sweden and Finland and, more recently, in other countries such as Canada, Poland and Czechoslovakia. Also, because machines were available to research laboratories, strength data has been accumulated much faster than ever before and on full size pieces.

To measure stiffness, machines can either employ the principle of measuring the force required to cause a fixed deflection, or to measure the deflection caused by a known force applied at a known point.

Stress grading machines operate with timber pieces fed continuously end-to-end. A typical design of machine used in Europe including the Nordic area has reaction rollers spaced 900 mm apart, takes measurements every time the timber moves forward 150 mm, and can operate at feed speeds of up to 90 metres per minute. The machines can deal with sawn or processed timber, dried or 'green'. The measurements of stiffness are processed by a computer which stores them until the piece clears the machine, calculates the strength based on the least-stiff part of the piece, and activates a marking device which marks the timber with the stress grade relevant to the least-stiff part of the timber. In addition, most machines are able to mark at 150 mm increments of length with a colour code related to the stiffness of the particular length which is central between the two reaction rollers, although this is not required by grading standards.

Because the stress grade is based on the weakest part of a length, a piece can be cut into two or more lengths without the stress grade of any length being reduced. If, however, the piece is resawn in cross-section, the stress grading is invalidated and each length must be regraded if to be used structurally.

There is a short length at the beginning and end of each piece which is not graded to the same extent as the rest of the length. However, unless there is a defect or characteristic within these short lengths which is much larger than any which occurs over the rest of the piece, experience has shown that the stress grade is unlikely to be affected.

It is known that as much as 70/80% of the strength of a piece can be determined by density. Machines take account of density, and therefore, if a piece has high density, the effect on strength can more than compensate for the effects of defects or characteristics such as knots. Consequently it is quite possible for knots larger than one would normally associate with visual stress grades to be passed by the machine. In cases where this has been questioned and the graded timber has been tested to destruction in a laboratory, the machine had been found to be correct in accepting the timber as being up to grade.

The machines being used commercially in the UK, and at Swedish, Finnish, Polish and Czechoslovakian sawmills to supply the UK, are administered under the BSI 'Kitemark' scheme and each piece of timber is marked accordingly. In North America, the tendency is to use different machines, different control procedures and different control bodies to those which have been used in the UK to match BS4978: 1988 by machine stress grading, and to grade only processed (surfaced) timber.

BS4978: 1988 describes two machine stress grades interchangeable with the visual stress grades 'General Structural' and 'Special Structural'. When a machine is set to select these grades, the timber is marked MGS and MSS. It is theoretically possible to set a machine to any level, although at present only two further machine stress grades are described in BS4978. For historical reasons these are given a numerical reference and identification M50 and M75. M75 is the strongest machine stress grade and, in the UK, it is considered to be too strong a grade to have a visually graded equivalent, mainly because currently there is no density check when timber is visually stress graded. M50 does not have a visual equivalent and is so close to MSS (although slightly weaker than MSS) that it is possibly unfortunate that it is retained. It is retained because it has been used in significant quantities in trussed rafter production. For comparison purposes, the dry grade stress in bending and the mean E value for the machine stress grades of BS4978: 1988 are tabulated in Table B3.1. As can be seen, in order of strength, they are:

MGS, M50, MSS and M75 (M75 being the strongest).

BS5268: Part 2: 1988 tabulates stresses for surfaced CLS North American machine stress graded timber. However, the pattern of supply is not yet well established and therefore it is not included in this book for comparison with the combinations in Tables B3.1 and B3.2.

Machines take account of most of the strength reducing characteristics of timber, but fissures, insect attack, wane and small amounts of decay have so little effect on stiffness that one cannot assume that a machine will detect them. BS4978: 1988 calls, therefore, for a certain amount of visual stress grading (sometimes referred to as a 'visual override') to limit fissures, insect attack, wane or decay.

The machines used in the UK, Sweden, Finland, Poland and Czechoslovakia are able to select two stress grades (plus reject) at one pass.

Although the relationship between stiffness and strength is adequate to enable commercial machines to be designed, it is not a 100% relationship. If a piece of timber which has a stiffness close to the indicating parameter of the machine (the 'go/no go' setting for a stress grade) is passed twice through a machine, it is possible therefore that one time it will (just) pass a grade level and the other time it will (just) fail. Such possibilities have been built into the factors of safety for machine stress grades. (For further reading on this point see BRE Information Sheet IS22/78.)

Machine settings are available for the laminating grades LA, LB and LC described in BS4978: 1988. It is possible, therefore, to use machines to select these laminates for glulam (see Chapter A11).

The maximum size of timber which can be stress graded by the machines used currently in the UK, Sweden, Finland, Poland and Czechoslovakia is 75 × 300 mm.

TOLERANCES ON SIZES

At one time it was a specific requirement of BS4978 that the tolerances on sawn and processed sizes had to be in accordance with BS4471 (i.e. virtually no minus tolerance permitted). This is still a requirement for machine stress graded timber but, for visually stress graded timber, the tolerances on size as given in BS4978: 1988 are in the form of a BSI 'note' which, although still referring to BS4471 tolerances, permits others to be used if specified. In view of this, if the specifier wishes to avoid the possibility of visually stress graded timber being supplied with minus tolerances different to those required, it is preferable for the specification to include a clause dealing with tolerances on size.

STRENGTH CLASSES

Particularly when one includes hardwoods and stress grades additional to those in BS4978, design code BS5268: Part 2 includes a multiplicity of stress grade/ species combinations, each with its own grade stresses. In an attempt to overcome the confusions which this variety causes to non-specialists, Strength Classes have been introduced *in addition to* stress grade/species combinations (which are retained). There are nine Strength Classes of which the lower five are intended to cover softwoods. The grade bending stress values in N/mm^2 for Strength Classes SC1–SC9 are:

SC1	SC2	SC3	SC4	SC5	SC6	SC7	SC8	SC9
2.8	4.1	5.3	7.5	10.0	12.5	15.0	17.5	20.5.

Four examples of how stress grade/species combinations match Strength Classes are:

GS	European Whitewood	matches Strength Class 3
M75	British-grown Sitka Spruce	matches Strength Class 3
SS	European Whitewood	matches Strength Class 4
M75	European Whitewood	matches Strength Class 5

A larger list of stress grade/species combinations which match SC2, SC3, SC4 and SC5 is given in Table B3.3.

Strength Classes can lead to simplification in the design of simple components such as floor joists and can be useful to a timber supplier. Strength Classes, however, can introduce their own complications where various species tabulated as satisfying a Strength Class or Classes have characteristics which are unacceptable for a particular requirement of the overall specification. Five possible complications are:

- If a designer specifies a Strength Class but wants the timber to have a prescribed durability or to be preserved, it will be necessary to limit the specification to species which have the required durability, or to require different species to be preserved by different preservation schedules. It is most unlikely that all the species which satisfy a particular Strength Class will have the same durability or the same amenability to preservation.

Table B3.3 A selection of stress grade/species combinations which match strength classes SC2, SC3, SC4 and SC5 of BS5268: Part 2: 1988

Species	Strength classes			
	SC2	SC3	SC4	SC5
European Whitewood		GS/MGS M50	SS/MSS	M75
European Redwood		GS/MGS M50	SS/MSS	M75
Canadian Spruce-Pine-Fir[1]		No. 1, No. 2 Joist & Plank	Select Joist & Plank	
		38 × 89 & 38 × 63 No. 1, No. 2 Struct L.F.	38 × 89 & 38 × 63 Select Struct L.F.	
	38 × 89 Const L.F.			
		GS/MGS M50	SS/MSS M75	
Canadian Hem-Fir[1] as Spruce-Pine-Fir except M75 matches SC5				
British-grown Sitka Spruce	SS/MSS M50	M75		

[1] Stresses given in BS5268: Part 2 are already increased (by 14.3%) by the width (depth) factor of BS5268: Part 2.

- There are certain species in Strength Classes SC3 and SC5 which must have joints designed as though the timber is in Strength Classes SC2 and SC4 respectively.
- Certain stress grade/species combinations of Canadian timber which are tabulated as satisfying Strength Class SC2 are given a zero rating in tension, therefore must be excluded from any SC2 specification for a member which is to take tension.
- In some cases, the sawn sizes and tolerances of North American timber and the 'basic' BS4471: 1988 sawn sizes and tolerances normally relevant to European timber are not interchangeable when the timber is dry, which is a point on which a supplier must be particularly careful when satisfying a Strength Class specification.
- If timber is to be finger jointed, any species mix must be excluded from a Strength Class specification because BS5291: 1984 on finger jointing of structural softwood does not permit different species to be finger jointed (even if each species in a mix would otherwise satisfy the strength requirements of the Strength Class).

If a designer or specifier decides to quote on the basis of stress grade/ species combinations, the specification should state which combinations are acceptable for the design. However, if the decision is to quote a Strength Class or Classes, the designer or specifier must make it clear if there are species which are not acceptable. Despite the apparent simplification of specifying a Strength Class, if the specifier is aware of the preferred species, there are advantages in quoting that species and the stress grade.

COMMON EUROPEAN STRESS GRADES

Research work carried on since 1968 as a collaboration between the forest products laboratories of Sweden, Finland, Canada and the UK represents much of the latest data on the strength of softwoods and how to relate grading to strength. It has been made available to the relevant laboratories of several European countries and has been accepted as the basis for drafting common European stress grades. These were first described in the ECE recommended standard 'Stress Grading of Structural Coniferous Sawn Timber', as S6, S8 and S10 (the figures approximating to the grade bending stresses) which has been proposed in current CEN work as one basis for establishing stress grades which satisfy Eurocode 5 (the Eurocode for timber).

S6 and S8 are interchangeable with the BS4978: 1988 stress grades General Structural and Special Structural, and this is acknowledged in BS5268: Part 2: 1988. There is no visual stress grade in BS4978: 1988 comparable to S10, but MS10 is at about the same level as M75.

NORDIC STRESS GRADED T-TIMBER

The Nordic area has had domestic stress grades known as T-Timber since the 1950s. However, the grades have not been sold to any great extent to the UK other than in kits exported for timber framed houses.

JOINERY GRADING WITH NOTES ON WOOD TRIM

B4

INTRODUCTION

Although a certain amount of special sawing and grading for joinery is carried out by a number of sawmills, particularly in Sweden, Finland and Canada, and although the amount of these specialities is increasing, the greatest volume of timber available to and purchased by UK joinery manufacturers is graded to one of the traditional sawmill grading rules (e.g. 'unsorted', 'fifth quality', 'Clears' etc.) *which are not end use grading rules and should not normally be quoted in end use joinery specifications.* As explained in the chapters in this book which cover sawmilling, normal mill grading rules are not particularly precise, being intended primarily as a means of trading between specialist sellers and buyers who have the opportunity to see and compare individual productions and, with experience, know the 'usual bracking' (i.e. the usual qualities) of the mills with which they deal.

It is important also to appreciate that re-sawing timber which has been graded at a sawmill into smaller pieces, as happens often in joinery production, *invalidates the initial grading.* Even when the timber is not to be resawn subsequently, except in special cases associated with the supply of specially sawn and graded material, a mill grader will not normally know which surfaces of sawn timber will be exposed or concealed in the final work, will not know the extent of any moulding which will reveal new surfaces, and will not know the paint or stain finish which is to be applied. Therefore, when a specifier wishes to define the quality of timber in a joinery item in a meaningful way with any degree of precision, *it is necessary to refer to an end use joinery standard.*

It is possible to spoil a perfectly good end use joinery specification by referring also to a sawmill grade. In such a case, reference to the sawmill grade can cause confusion or can even contradict the valid part of the specification, because it is not normally possible for the quality of timber to comply with two separate grading rules. If a specifier insists on referring to a sawmill grading rule as well as to an end use grading rule, care must be taken with the wording to make it clear that the timber is to be selected *from* the timber as graded at the sawmill, and regraded *to* the end use grade. Normally, however, it is better not to refer at all to a sawmill grade in an end use specification.

The principal UK joinery standards are BS1186 Part 1: 1986 on timber in joinery, and Part 2: 1987 on workmanship in joinery. BS1186: Part 1 was amended considerably in the 1986 version as was the 1987 version of Part 2. It is important to realize that BS1186: Part 1: 1986 is a base standard and therefore may contain clauses which are not strictly applicable to all joinery components, and hence may be excluded or amended in a product standard or by a specifier. It may be necessary or advantageous to add one or more clauses when specifying for a specific or unusual application.

When specifying, it is meaningless simply to state 'to BS1186: Part 1'. It is necessary to refer to the relevant Class or Classes of BS1186, as described below in the notes on BS1186: Part 1: 1986.

From the 1986 version of BS1186: Part 1, it is clear that it is not the intention that BS1186: Part 1 should cover the quality of timber in exterior wood trim such as architraves, barge boards and timber cladding, or interior wood trim such as architraves and interior boarding. The intention is that the specification of such items will be covered by a new standard BS1186: Part 3 *Timber for and workmanship in joinery. Specification for wood trim: timber species, classification, workmanship in fixing, and standard profiles*, which has passed the BSI stage of 'public comment'. When published, the intention is that it will replace BS584: 1967 (1980) *Specification for wood trim (softwood)*. Although this chapter is devoted mainly to joinery, some notes are added on wood trim based on the assumption that BS1186: Part 3 will be published generally as drafted at the time of 'public comment', but with some amendments made in 1989 on the basis of 'public comment' before the draft was released for publication by BSI (probably in 1990). Readers are referred, however, to the actual standard when published. Also see Chapters A4 and A5 for further notes on 'wood trim' as intended by the rather wide definition in BS1186: Part 3. The clauses in BS1186: Part 3 which deal with timber quality are based on a simplified version of BS1186: Part 1: 1986. One example of this simplification is that Part 3 sets hardly any requirements for a 'Concealed' surface. Until BS1186: Part 3 is published, a specifier of a written specification has litle choice but to prepare an individual detailed specification (perhaps based on the draft of BS1186: Part 3), or to refer to BS1186: Part 1: 1986 (perhaps relaxing certain clauses in line with the draft of BS1186: Part 3).

Note that the term 'framed panelling' as used in Appendix B of BS1186: Part 1: 1986 is intended to apply to solid timber around panels of plywood etc., and not to the type of tongued and grooved boarding such as 'knotty pine' etc., which is intended to be covered in BS1186: Part 3 when published.

Generally one should not over-specify. If, for example, all that is required for wood trim is a straight moulded section with no wane, then the specifier should consider just calling for that.

JOINERY SOFTWOODS

The softwoods covered by this book most likely to be used for joinery (and wood trim) are:

European Redwood

BS1186: Part 1: 1986 (and the draft of Part 3 for wood trim) calls for this species to be preserved when used for exterior joinery in the UK. It is relatively easy to preserve to the 30 or 60 year desired service life performance category of BS5589: 1989, and easy to process. Care should be taken to avoid or minimize blue stain by keeping the timber dry, particularly if a non-opaque finish is to be used. The normal sources of supply of joinery quality European Redwood are Sweden, Finland and Russia.

European Whitewood

BS1186: Part 1: 1986 (and the draft of Part 3 for wood trim) calls for this species to be preserved when used for exterior joinery in the UK. It is not as easy to preserve as European Redwood, but it can be preserved commercially for exterior joinery with an organic solvent to the 30 year desired service life performance category of BS5589: 1989, and by a water borne process to the 60 year category. European Whitewood has a natural property of shedding water to a remarkable extent which makes it particularly suitable for use, for example, as exterior cladding. Care must be taken when machining to maintain the correct moisture content and the sharpness of cutters, and to use cutter angles suited to the moisture content. For joinery, it is not normal in the UK to purchase Whitewood from mid-European sources which includes a percentage of 'Silver Fir'. The normal sources of supply of joinery quality European Whitewood are Sweden, Finland and Russia.

Canadian Douglas Fir-Larch

This species can normally be used for exterior joinery (or exterior wood trim) in the UK without being preserved, the assumption being that no sapwood is present. The prominent growth ring 'figure' can show through the finish even if an opaque paint is used. The species is stated in BS1186: Part 1: 1986 (and the draft of Part 3) as being susceptible to iron staining if used in damp conditions with ferrous fasteners.

Canadian Hemlock

BS1186: Part 1: 1986 (and the draft of Part 3 for wood trim) calls for this species to be preserved when used for exterior joinery in the UK. For most exterior uses it can be preserved with an organic solvent to the 30 year desired service life performance category of BS5589: 1989 but, for exterior doors, it can be preserved with an organic solvent to the 60 year category. (See Chapter B5 for the explanation of this point.) By using a water borne process, it can be preserved to the 60 year category for exterior joinery or wood trim. Shipments may contain a proportion of Amabilis Fir. If preserved by an organic solvent, the Amabilis Fir may absorb considerably more solvent than the Hemlock. Care is needed to obtain a clean cut when cutting across the grain, or to prevent splitting when nailing.

Canadian Western Red Cedar

This species can normally be used for exterior joinery in the UK without being preserved. It is stated in BS1186: Part 1: 1986 (and the draft of Part 3) to be susceptible to iron staining if used in damp conditions with ferrous fasteners, and there is a tendency for corrosion of ferrous fasteners if used in contact with this species in damp conditions.

Parana Pine

This species is not considered in BS1186: Part 1: 1986 (or in the draft of Part 3) to be suitable for exterior joinery in the UK. It is a fine looking timber for interior joinery with interesting colour variations. Due to a tendency to distort on drying, its main uses are those where sections are restrained at close centres (e.g. stair strings and treads, window boards, drawer sides etc.). The source of supply for the UK is Brazil.

BS1186: Part 1: 1986 does not rate any of the softwoods listed above as being suitable for door sills or thresholds (unless, of course, given a protective cover).

See Appendix B of BS1186: Part 1: 1986 (and Appendix B of the draft of BS1186: Part 3) for further guidance on the uses of the six softwoods listed above.

THE QUALITY OF TIMBER IN JOINERY TO SATISFY BS1186: PART 1: 1986

BS1186: Part 1: 1986 defines four 'Classes' for the quality of timber in joinery (as does the draft of BS1186: Part 3 for wood trim). These are denominated:

Class 3,
Class 2,
Class 1,
Class CSH.

The abbreviation CSH refers to 'Clear' grades of Softwoods and 'Clear' grades of Hardwoods.

The clauses in BS1186: Part 1: 1986 are intended to cover the quality of timber at the time that the joinery is handed over by the joinery manufacturer to the first purchaser. The quality of wood trim in Part 3 as drafted is either at the time of handover or after fixing, whichever is relevant.

The clauses are written for cases where an opaque paint is to be used. Where, however, a non-opaque finish is to be used, the clauses as drafted are still intended to apply unless specifically varied by the specifier or manufacturer. The attention of the specifier is drawn to certain clauses: i.e. on splits, shakes and checks, unsound knots, resin pockets, discoloured sapwood, insect attack, plugs, inserts and fillers, laminating, edge jointing and finger jointing, which may have to be varied if a non-opaque finish is to be used. Because of the multiplicity of non-opaque finishes which are available, the varying degree to which each will conceal characteristics or 'making good', and varying opinions amongst specifiers as to what is or is not an acceptable appearance, it is necessary for the final decision on varying or not varying selected clauses to be left to individual specifiers once a finish has been chosen.

The point is made in BS1186: Part 1: 1986 (and in Part 3 as drafted) that, if a specifier does not vary any clauses, the joinery manufacturer is entitled to assume that the clauses as they appear in Part 1 are acceptable, even if a non-opaque finish is to be used. The point is also made that varying a clause may lead to an increase in cost and to an extended delivery period.

The notes below discuss some of the important points in BS1186: Part 1: 1986 but it is such an important standard that specifiers are encouraged to obtain a copy and to study it in detail. Part 3 as drafted has similar clauses.

Classes 2 and 3

Classes 2 and 3 are intended to cover most general-purpose 'normal quality' joinery items made from softwoods such as European Redwood and European Whitewood. Classes 2 and 3 can refer to solid, laminated or edge jointed sections.

Class 1

Class 1 is intended to cover high quality or specialist joinery items where the cost of the additional selection of material (which can be quite considerable) is acceptable to the purchaser. The basic material from which Class 1 is likely to be selected is unlikely to be held in stock in several species in all sizes by an importer or joinery manufacturer, and this may well lead to an extended delivery period. Class 1 can refer to solid, laminated or edge jointed sections.

Class CSH

As far as the softwoods covered by this book are concerned, with the exception of small or short pieces, Class CSH is likely to be satisfied only by using 'Clear' grades of Douglas Fir-Larch, 'Clear' grades of Hemlock (or clear pieces cut between knots), Parana Pine or Western Red Cedar, and by accepting the cost, which may be considerable.

Class CSH will be satisfied by European Redwood or European Whitewood only in small short pieces such as glazing beads, and then only by careful selection. Specifiers should not normally call for Class CSH for joinery or wood trim which uses larger sizes of European Redwood or European Whitewood.

Class CSH can refer to solid, laminated or edge jointed sections.

Specifying a mix of Classes

It is permitted and quite common to specify a mix of Classes in one joinery item. For example, NHBC and BS644: Part 1: 1989 on wood windows call for the timber in window casements (see Fig. A1.1) to match Class 2, and the timber in the main frames to match Class 3.

Surface categories

BS1186: Part 1: 1986 distinguishes between the requirements of:

- an '*Exposed Surface*';
- a '*Concealed Surface*', i.e. one which is not 'Exposed' or 'Semi-Concealed' in the final construction;
- a '*Semi-Concealed Surface*' i.e. one which is not usually 'Exposed' when the item is in the closed position but which becomes 'Exposed' when the item is in the open position, as with some surfaces of a window.

Several of the clauses in the draft of BS1186: Part 3 are a simplified version or less onerous than the corresponding requirements of Part 1: 1986, particularly for 'Concealed' surfaces. In the draft of Part 3, there is no mention of 'Semi-Concealed' surfaces.

It is the intention of the Parts of BS1186 that an 'Exposed' surface is still considered to be 'Exposed' even if covered by an opaque paint.

Different limits of knot sizes, fissures etc. are given for each category of surface. In detailing the clauses on 'making good', BS1186: Part 1: 1986 covers the 'before' condition as well as the 'after' condition. This is perhaps unnecessary because, for example, if a plug or filler of a certain size is acceptable on the final surface, it hardly seems to matter what size of defect it has replaced.

Guidance is given on the moisture content requirement of joinery at the time of handover by the joinery manufacturer to the first purchaser. This should be close to the average moisture content which the item will attain in service (see Table B2.2 in Chapter B2). The clauses on moisture content in the draft of Part 3 on wood trim are more simple than those in Part 1. See Table A4.1 and Chapters A4 and A15 of this book and BS1186: Part 3 when published.

Knots

BS1186: Part 1: 1986 makes the point that 'whilst sound knots, tight knots and knot clusters are not detrimental to the use of timber in joinery, it is desirable to limit both their size and distribution'. This can be for appearance or to ensure uniformity, or to ensure that quotations from different manufacturers are produced on the same basis. The graphs of permissible knot sizes given in BS1186: Part 1: 1986 include a diagrammatical way of presenting permissible deviations related to knot distribution. Specifiers should be aware of these. For simplicity, however, Fig. B4.1 gives graphs for what could be regarded as the 'target line' for round knots for graders to work to. These are given for 'Exposed' and 'Semi-Concealed' surfaces of each Class, and one graph is added which is applicable to 'Concealed' surfaces of all four Classes. The simplified version of the knot size limitations given in Fig. B4.1 is

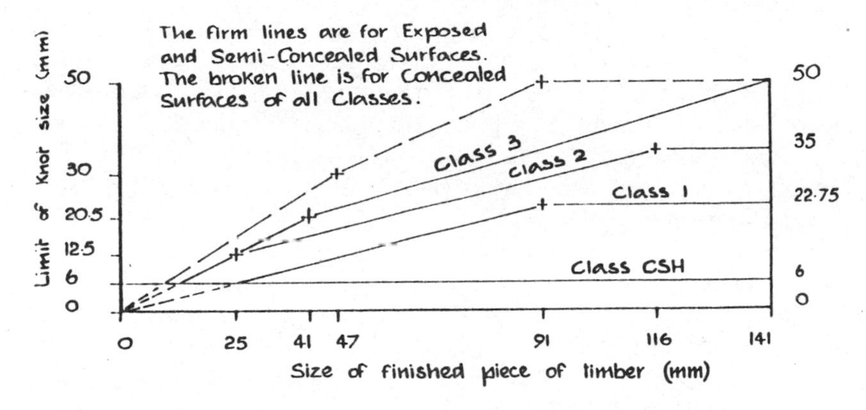

Fig. B4.1 'Target line' limits for round knots to comply with BS1186: Part 1: 1986

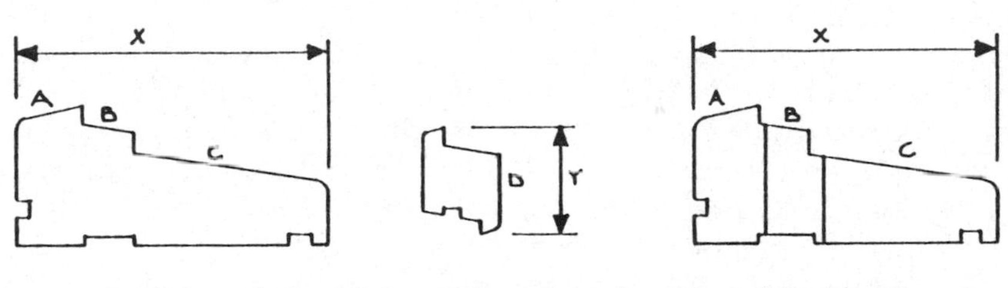

Fig. B4.2 Dimensions on which knots are assessed

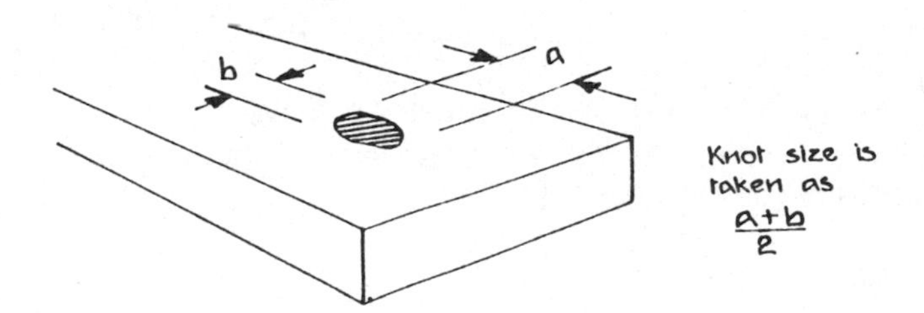

Fig. B4.3 Method of measuring round and oval knots

the version adopted in the draft of BS1186: Part 3 for 'Exposed' surfaces.

Figure B4.2 illustrates the dimensions of the surface on which a knot occurs which are the dimensions to consider when determining if the size of a knot is within the limit for the Class. Note that it is not always the most obvious dimension. For example, it can sometimes be continuous on both sides of glass in a window.

The method of measuring a round or oval knot for the purpose of comparison with the limit on knot size in BS1186: Part 1: 1986 for a particular Class and timber dimension is illustrated in Fig. B4.3. The method of measuring other types of knot is given in BS1186: Part 1: 1986 (and repeated in the draft of Part 3). The size thus established is then compared to the size limit of a round knot for the particular Class and timber dimension (see Fig. B4.1).

Unsound knots, dead knots and loose knots are acceptable only on Concealed surfaces. If they occur on an Exposed surface they may be made good by a plug, insert or filler if permitted for the Class and not specifically excluded by the specifier. The size of a plug or insert is limited (see the notes below on 'Plugs and Inserts').

Type of knot

Experience shows that often there is confusion about whether or not a knot is 'live' or 'dead'. Also, discussions take place about what causes the dark ring around some knots, and why there is resin exudation from some knots. The phrase 'live knot' is used regularly but it is worth noting that neither BS6100: Section 4.1: 1984 nor BS1186: Part 1: 1986 contain a definition of a 'live knot', although Definition 410 5016 in BS6100: Section 4.1: 1984 does give a definition of an 'intergrown knot' ('live knot' definition being non-preferred) as being a:

> 'knot having fibres intergrown with those of the surrounding wood to the extent of approximately 75% or more of the cross-sectional perimeter.' (Presumably 'cross-sectional perimeter' is intended to refer to the total periphery of the knot within the thickness of the timber.)

If the words 'live knot' are used, it seems reasonable to accept that the BS definition of an 'intergrown knot' is what is intended.

BS1186: Part 1: 1986 refers in particular to

'sound knots',
'dead knots',
'tight knots',
'unsound knots' and
'loose knots'.

As defined in BS6100: Section 4.1: 1984:

- A '*sound knot*' is a knot free from rot, solid across its face and at least as hard as the surrounding wood.

- A '*dead knot*' is a knot having fibres intergrown with those of the surrounding wood to the extent of 25% or less of the cross-sectional perimeter.
- A '*tight knot*' is a knot held firmly in place. (Note that there is no requirement for the extent to which the fibres of a 'tight' knot must be intergrown with the surrounding wood, therefore a 'dead' knot can be described as a 'tight' knot.)
- An '*unsound knot*' is a knot not solid across its face, or one which is softer than the surrounding wood due to rot or other defects.
- A '*loose knot*' is a dead knot that is not held firmly in place.

Reference to Fig. B4.4 and the following notes should help to explain how to consider and define a knot by understanding the 'history' of the branch which led to the knot, how a knot appears on the surface of

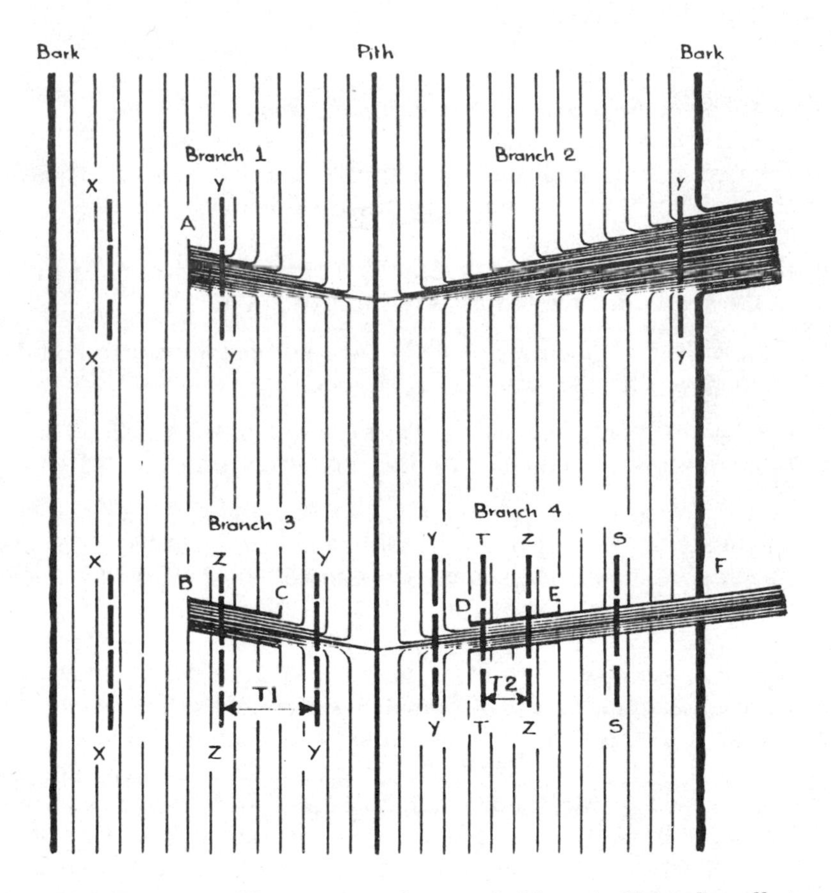

Fig. B4.4 Diagram to illustrate how the growth 'history' of branches affects the way in which knots appear in sawn timber

timber converted from the tree, and if a knot is 'intergrown', 'dead', 'tight' or 'loose' within the thickness of the timber.

Branches of the growing tree shown in Fig. B4.4

Branch 1 was pruned at A, close to the bark, at the time when the tree had grown outwards as far as A.

Branch 2 is still growing and has not been pruned.

Branch 3 died when the tree had grown outwards as far as C, after which the dead branch and the attached bark was occluded by the growing tree until the dead branch was pruned or broken-off at B.

Branch 4 died at the time when the tree had grown outwards as far as D. Bark remained on the branch from D to E after which point the bark became detached. The branch was not pruned and continued to be occluded by the growing tree.

Knots in the sawn timber sawn from the tree shown in Fig. B4.4

If the tree is converted into sawn timber, the surfaces of the sawn timber will have the appearance as described below:

On each surface identified as XX, the timber will appear to be 'knot-free' even though the timber is not free of knots.

On each surface identified as YY, the knot will appear to be fully intergrown with the surrounding timber.

On the surface identified as ZZ the knot will appear as being 'bark-ringed' (or 'encased' to use the preferred BS6100: 4.1: 1984 definition).

On the surface identified as SS, the knot will appear as having a dark ring around it (which is sometimes assumed, mistakenly, to be due to a kilning defect).

If a piece of timber sawn from the tree has a thickness T1 between Surface YY and Surface ZZ of Branch 3 (see Fig. B4.4), on Surface YY the knot will appear to be fully 'intergrown' (and is likely to be described as being 'live' by most non-specialists in timber—and by many specialists) whilst, on Surface ZZ, it will appear as being 'encased' by bark. The knot is sufficiently intergrown over part of its length in the timber to satisfy the definition of being 'tight' no matter from what surface it is viewed. As sketched, it is 'sound', not 'dead'.

If a piece of timber sawn from the tree has a thickness T2 between

Surface ZZ and Surface TT of Branch 4 (see Fig. B4.4), the knot will appear as being 'encased' by bark on both surfaces. It is not intergrown at all with the surrounding timber and therefore can be described as 'dead'. It is 'loose'. It may be so loose that it will drop out, in which case there will be a knot hole.

Resin

Softwoods develop resin at places where the growing tree is damaged by 'natural' or other causes. (Hardwoods develop 'gum'.) Referring to Fig. B4.4, between C and B, and between D and E, resin is likely to occur as well as bark. Between E and F, resin is likely to occur as well as a 'dark ring' (i.e. a dark ring rather than bark). On Surfaces ZZ, TT and SS of the sawn timber, the ring around the knot will not have been caused by kilning as is sometimes assumed; it is due to the growth history of the branch. However, the temperature during kiln drying or subsequent heat in service (say under a dark stain) may cause resin to exude around the knot and appear on the surface of the sawn timber.

Some timbers (e.g. European Whitewood) have resin pockets in a small number of pieces. These occur in a random way, possibly caused as a result of a combination of frost and wind affecting the growing tree.

Splits, shakes and checks

A split is a separation of the fibres which goes right through the piece between two surfaces. Except for short end splits, they are not permitted in joinery which has to comply with BS1186: Part 1: 1986. The draft of BS1186: Part 3 permits short end splits at the time of delivery, provided that they are cut off before the wood trim is fixed in place. A ring shake (i.e. a shake which follows the line of a growth ring) must not occur on any 'Exposed' or 'Semi-Concealed' surface in any BS1186 Class, and not on any surface of Class 1 or Class CSH.

A shake or check is a separation of the fibres which does not penetrate to another surface. Shakes or checks are permitted to a greater or lesser extent as detailed in BS1186: Part 1: 1986 and the draft of Part 3. The depth of a shake or check is judged by inserting a feeler gauge not exceeding a thickness of 0.2 mm.

Experience in practice shows that a shake or check of 0.5 mm width or less cannot be filled. Where a non-opaque finish is to be used, the specifier must decide (perhaps with the paint manufacturer) if a filler of adequate durability is available and if an adequate colour match can be achieved under the finish. With many decorative stains, if a filler is used, it can actually accentuate the presence of a shake or check rather than conceal it, because of the different absorbency and colour of the surface of the filler compared to the surrounding timber.

Resin pockets

If permitted by the Class limits, and if suitable for the finish, resin pockets which occur on an 'Exposed' or 'Semi-Concealed' surface may be removed and made good by a plug, insert or filler. Resin pockets are permitted to occur on a Concealed surface of any Class of BS1186: Part 1: 1986 without having to be made good.

Sapwood and discoloured sapwood

Sapwood is permitted. Exterior joinery timber containing sapwood must be preserved to satisfy BS1186: Part 1: 1986. Likewise, exterior wood trim containing sapwood must be preserved to satisfy the draft of BS1186: Part 3.

Discoloured sapwood such as blue stain is not a defect which affects the structure of the timber. It will not spread at moisture contents of less than about 20/22%. Discoloured sapwood is permitted by BS1186: Part 1: 1986 and the draft of Part 3 to a limited extent unless it would adversely affect the appearance of a non-opaque finish. If discoloured sapwood is not permitted (e.g. because of the finish which is to be applied), the specification must make this clear.

Wane

Wane will not normally be permitted nor will it normally occur in finished joinery or wood trim. If it does occur, it should not be visible in the final construction except in the rare cases where it is called for as a special feature (e.g. waney-edged cladding).

Rate of growth

BS1186: Part 1: 1986 specifies limits for the rate of growth of softwoods. At least an average of 6 growth rings per 25 mm are required for joinery for exterior use, and at least an average of 4 growth rings per 25 mm are required for joinery for interior use. The same requirements appear in the draft of BS1186: Part 3. It is doubtful if limits of growth are particularly meaningful in a base standard such as BS1186 for all joinery and wood trim, other than to set an absolute minimum requirement. The main intention of the limits is to exclude timber having extremely wide growth rings. Product standards may well set tighter limits.

Straightness of grain

As with rate of growth, it is doubtful if limits on slope of grain are entirely meaningful in a base standard such as BS1186 other than to set a minimum requirement. For softwoods used in an exterior or interior situation, BS1186: Part 1: 1986 and the draft of Part 3 call for the slope of grain not to exceed 1 in 10. This may not be tight enough for the stiles of a panel door, but may be unnecessarily tight for short, well-restrained members. Slope of grain is intended to be a measure of the general direction of the grain not taking account of local deviations.

Pith

Pith is permitted by BS1186: Part 1: 1986 only on 'Concealed' and 'Semi-Concealed' surfaces (however, presumably if there is a short length of pith on an 'Exposed' surface, it may be cut out and be made good if the extent of the making good is permitted by the Class limits and if suitable for the finish). In the draft of BS1186: Part 3 as released, in 1989, for printing by BSI, pith *is* permitted on 'Exposed' surfaces of exterior cladding and interior boarding of Classes 1, 2 and 3 (but not Class CSH).

Decay and insect damage

As far as insect damage is concerned, BS1186: Part 1: 1986 and the draft of Part 3 permit only Ambrosia beetle damage (sometimes referred to as pinhole borer damage). Only a small amount is permitted and then only if the holes on an 'Exposed' or 'Semi-Concealed' surface can be filled without detriment to the intended finish. Decay is not permitted at all.

Plugs and inserts

Plugs and inserts are permitted by BS1186 on 'Exposed' and 'Semi-Concealed' surfaces of Classes 1, 2 and 3. Plugs must be of the same species as the surrounding timber, occupy the whole depth of the hole, have the grain in the same general direction as the surrounding timber, and be secured by a suitable adhesive of adequate durability. The lesser surface dimension of a plug must not be more than 6 mm greater than the maximum size of knot permitted for the section size and Class. A plug can be round, oval or specially shaped. Two plugs (usually round) can be overlapped. If the specifier decides that, due to a non-opaque finish being used, plugs and inserts should not be used, the joinery manufacturer must be informed in good time.

Filler

Filler, if used, must have adequate durability for the service conditions. The size of any individual hole to be filled is limited by BS1186: Part 1: 1986 and the draft of Part 3 as for plugs and inserts. The limit on the filling of shakes and checks is detailed in BS1186 for the various Classes. Experience in practice shows that a shake or check of 0.5 mm width or less cannot be filled. If a non-opaque finish is to be used, the specifier must decide if an acceptable filler of suitable durability, absorbency and acceptable colour match is available and, if so, decide (preferably with the joinery manufacturer and the manufacturer of the finish) which type and colour is to be used. It is worth repeating that, if a filler is used under a non-opaque finish, it may actually accentuate the presence of a fissure rather than conceal it due to the different absorbency and colour of the filler compared to the surrounding timber.

Laminating or edge jointing

Laminating or edge jointing of joinery sections is becoming more common and favoured in the UK. If properly carried out, laminating adds considerable stability to a section. Obviously a suitable adhesive of adequate durability must be used (see Chapter B9). Laminating or edge jointing is permitted by BS1186 including in sections which are to receive a non-opaque finish.

End jointing is usually an integral part of laminating, therefore each laminate may well contain finger joints or butt joints. Finger joints for joinery laminating should be of a type with no apparent gap at the finger tips (see Fig. B10.1).

The size of knot (or part of a knot) permitted on the surface of a laminated section is set by BS1186 in relation to the overall dimension of the surface on which the knot occurs, not by the size of individual laminates (see Fig. B4.2).

Finger jointing

Finger jointing of solid sections or laminates is permitted by BS1186 providing that the profile of the joint is one with no apparent gap at the finger tips (see Fig. B10.1). A suitable adhesive of adequate durability must be used. If a non-opaque finish is to be used, and if the specifier decides that finger jointing is not acceptable, this must be stated, otherwise the manufacturer is permitted by BS1186 to assume that it is acceptable. Note that, if laminating is to be used, the manufacturer will normally assume that finger jointing (or butt jointing) of individual laminates is acceptable, because end jointing is usually an integral part of laminating.

SPECIES

Appendix B of BS1186: Part 1: 1986 (and Appendix B of the draft of BS1186: Part 3) gives guidance on the suitability for typical exterior and interior applications of six of the species covered by this book. They are:

European Redwood
European Whitewood
Western Hemlock
Douglas Fir-Larch
Western Red Cedar
Parana Pine.

One significant change from the 1971 version of BS1186: Part 1 is that European Whitewood is shown for most exterior uses as being 'SP' (suitable if preserved) where previously it was shown as being 'X' (unsuitable). It is now accepted in the UK that European Whitewood can be preserved commercially to acceptable levels for most items of exterior joinery and wood trim.

B5 DURABILITY AND PRESERVATION

INTRODUCTION

Each species of timber has a certain natural durability which is normally greater, and never less for the heartwood than the sapwood. Durability in this context is a measure of the natural resistance to fungal decay, not to insect attack. The Princes Risborough Laboratory of the Building Research Establishment (BRE) evolved a classification of durability which is shown in Table B5.1. It is important to realize, however, that it is possible to refer to durability only in relative and indicative terms, and to note that the classification refers to the performance of 50.8 × 50.8 mm (2 inch × 2 inch) test pieces of heartwood 'staked' in the ground. By reference to Table B5.1, it can be seen that the heartwood of timber classified as 'very durable' can be expected to have an approximate life in excess of 25 years even if in contact with the ground. In general, larger pieces would have longer life.

It is also important to realize that even species with low natural durability will not decay if the moisture content in service is kept below 20/22%. Even a piece of timber given the rather emotive description of 'perishable' will not decay if the details and service conditions cause it to have a moisture content in service of 20/22% or less. Even occasional short periods when the moisture content is increased above 20/22% will not lead to decay occurring, although discoloration (e.g. 'blue stain') may occur in the sapwood of some species if the moisture content is maintained above 20/22% for a few weeks.

Table B5.1 BRE classification for durability

Grade of durability as evolved by BRE	Approximate life (years) in contact with the ground for stakes of 50.8 × 50.8 mm sections
Very durable	More than 25
Durable	15–25
Moderately durable	10–15
Non-durable	5–10
Perishable	Less than 5

Table B5.2 Durability categories

Timber in ground contact	Grade of durability of:	
	Unpreserved heartwood	**Unpreserved sapwood**
European Redwood	Non-durable	Not known to the same extent as heartwood but probably close to or better than the boundary between non-durable and perishable
European Whitewood	Non-durable	
Hemlock or Hem-Fir	Non-durable	
Spruce-Pine-Fir	Non-durable	
Douglas Fir-Larch	Moderately durable	
Western Red Cedar	Durable	
Maritime Pine	Moderately durable	
British-grown Sitka Spruce	Non-durable	
Parana Pine	Non-durable	

On the other hand, it is as well to realize that, even if a species with high natural durability, or correctly preserved timber is used where moisture is trapped (for example by a poor detail or the use of an unsuitable or badly maintained paint), although the timber will last for a satisfactory period of time, the trapped moisture may cause trouble in another way (e.g. it may cause paint to peel off). The use of durable timber or preserved timber should never be considered to be an alternative to good detailing.

The sapwood of almost all timbers, not just softwoods, is either 'perishable' or 'non-durable'. (Specifiers should not fall into the trap of assuming that all hardwoods are durable.) Table B5.2 lists the durability category of the heartwood and sapwood of the softwoods covered by this book, based on information extracted mainly from *A handbook of softwoods*, published by the Building Research Establishment.

In Table B5.2 the heartwood of Douglas Fir-Larch is shown as being 'moderately durable' and that of Western Red Cedar as being 'durable'. In deciding whether or not to preserve these species, it is normal practice to consider that little if any sapwood will be present. In considering Maritime Pine, however, for which the heartwood is rated as 'moderately durable', it is advisable to take note of *A handbook of softwoods* by BRE in which it is indicated that a large part of this fast grown timber comprises sapwood.

AMENABILITY TO PRESERVATIVE TREATMENT

The two principal types of preservatives used in the UK for new building components are:

- water borne salt types of which the most common are normally referred to as copper/chromium/arsenic, or simply CCA, which are applied by pressure;

- organic solvent types which can be applied by a double vacuum process or by immersion.

They are described in detail later in this chapter.

The Princes Risborough Laboratory of BRE evolved the following classifications (before the use of organic solvents became commonplace) to indicate the amenability of species to preservative treatment.

Permeable

These timbers can be penetrated completely under pressure without difficulty, and can usually be heavily impregnated by the open tank process.

Moderately resistant

These timbers are fairly easy to treat and it is usually possible to obtain a lateral penetration of the order of 6–19 mm in about 2–3 hours under pressure.

Resistant

These timbers are difficult to impregnate under pressure and require a long period of treatment. It is often very difficult to penetrate them laterally more than about 3–6 mm. Incising is often used to obtain a better treatment. (Although the definition in *A handbook of softwoods* by PRL/BRE states that incising is 'often used', this is not the current case in the UK where it is rarely used for building timbers.)

Extremely resistant

These timbers absorb only a small amount of preservative even under long pressure treatments. They cannot be penetrated to an appreciable depth laterally and only to a very small extent longitudinally.

Amenability of species to preservative treatment

The amenability to preservative treatment of the species covered by this book is shown in Table B5.3. Note that the sapwood of a particular species is not always more permeable than the heartwood. Since the

Table B5.3 Amenability of species to preservative treatment

Species	Amenability to treatment of:	
	Heartwood	Sapwood
European Whitewood	Resistant	Resistant
European Redwood	Moderately resistant	Permeable
Canadian Douglas Fir-Larch	Resistant	Moderately resistant
Canadian Hemlock[2]	Resistant	Data not available
Spruce-Pine-Fir[1]	Resistant[1]	Data not available
Western Red Cedar	Resistant	Resistant
Parana Pine	Moderately resistant	Permeable
Maritime Pine	Resistant	Permeable
British-grown Sitka Spruce	Resistant	Data not available

[1] Amenability quoted for the Spruce.

[2] Note that BS5589: 1989 *Code of practice for preservation of timber*, permits Hemlock to be considered as being 'moderately resistant' when being preserved for exterior doors, but requires it to be taken as 'resistant' when being treated for other exterior joinery. Although this seems to be illogical, the explanation given is that tests on Hemlock door stock showed it to be less resistant to preservation than other Hemlock supply.

classifications described above were evolved, double vacuum organic solvent treatments and more advanced schedules for water borne preservatives have been developed, by which 'resistant' species can be treated commercially to acceptable levels for many uses. Some 'resistant' species (e.g. European Whitewood) are treated in the UK in significant quantities.

RISK AND AVOIDANCE OF FUNGAL DECAY

As stated above, all codes and standards are in agreement that there is little risk of fungal decay, even in sapwood, if the timber is maintained at a moisture content in service of 20/22% or less, therefore, for most cases of timber used inside a building there is rarely a 100% case for the use of preservation against decay unless there is a local hazard. Also, as shown by a study in particular of Table 4 of BS5268: Part 5: 1989 on the preservative treatment of structural timber, if one could ensure that only heartwood would be used, there would be even less need to preserve timber in building. In the notes below, however, which discuss the schedules to use, the assumption is that a decision has been taken by the specifier to preserve. Also see the 'Risk categories' in Tables B5.4 and B5.5.

For the purpose of interpreting 'dry' and 'wet' conditions as used in Table 3 of BS5268: Part 5: 1989, it should be acceptable to consider a 'dry' condition as one leading to a moisture content in the timber of 'below about 20%' (see Clause 3.2a of BS5268: Part 5: 1989).

Even for many exterior cases of joinery or carpentry, any risk of decay which may exist can be minimized by good practice and detailing. Over recent years, however, it has become quite common to preserve many components where, strictly speaking, the risk of decay does not warrant it or hardly warrants it. This has come about for one or more of several reasons, such as:

- a reaction to problems caused in the past by the use of a poor detail or an unsuitable paint, even once the poor detail or the unsuitable paint is no longer used;
- a concern about the possibility of condensation occurring within a structure, even for conditions where experience in practice shows that this does not happen;
- writers of regulations being over-cautious;
- quality assurance organizations calling for preservation to reduce their risk;
- builders, manufacturers or preservative companies promoting preservation as a sales advantage.

In deciding whether or not to call for preservation, the specifier is faced with several codes, standards and regulations to consider. Some of these are written in general terms whilst others are specific. The important ones are discussed later in this chapter to give guidance on their content, and are also referred to in the chapters in this book (i.e. Chapter A1, A2 etc.) which deal with specific components.

RISK AND AVOIDANCE OF INSECT ATTACK

Reference is made below to the four most common forms of insect attack encountered in the UK (and to termites). Also see Tables B5.4 and B5.5.

Ambrosia beetles (Pinhole borer)

The standing trees or recently felled logs of several softwoods can be attacked by Pinhole borer. The attack ceases and dies when the timber is converted and dried. Attack is more common in sapwood than heartwood. The defect in timber takes the form of circular holes or short tunnels of 0.5–3 mm diameter mainly across the grain. The holes are dark stained and contain no bore-dust.

*House Longhorn Beetles (*Hylotrupes bajulus L.*)*

There is one area of the Home Counties of England which is defined in Table 1 of the Approved Document to support Regulation 7 of the Building Regulations for England and Wales, where House Longhorn Beetles are a risk to softwood in roof voids of pitched or flat roofs where there is sufficient warmth. They can attack dry softwood.

*Furniture beetles (*Anobium punctatum*)*

The Furniture beetle (sometimes referred to as the 'common Furniture beetle') is the most common form of insect attack in the UK. Furniture beetles can attack dried sapwood, and even dried heartwood.

*Wood-wasps (*Siricidae*)*

Wood-wasps attack standing trees, and logs. The attack ceases and dies when the timber is converted and dried. The tunnels are circular and are filled with tightly-packed bore-dust.

*Termites (*Isoptera*)*

Termites are not a hazard in the UK and are not discussed in this book. Where a specifier is involved with an export contract to an area where termites exist, reference should be made to BS5589: 1989 which contains useful information on preserving against termites.

Treatment against insect attack

If it is decided to treat against insect attack by an organic solvent type preservative, the specifier should ensure that the formulation is one which contains an insecticide. Most of the organic solvent preservatives used for preserving softwoods for construction timbers contain an insecticide, but most used to preserve exterior softwood joinery do not contain one because normally there is no risk in the UK of insect attack in softwoods in exterior joinery. Note that CCA preservatives used with the correct schedules are effective against the insects encountered in the UK.

In the area of the Home Counties where House Longhorn Beetles are a risk, the Approved Document to support Regulation 7 of The Building Regulations for England and Wales requires softwood in 'roof voids' to be 'adequately treated'. Examples of adequate treatments would be those taken from BS5268: Part 5: 1989 as being suitable for use against House Longhorn Beetles.

It is not normally necessary for a specifier to go to the extent of specifying the actual preservation schedules which are permitted to be used when preserving timber in buildings against insect attack. However, if a check has to be made to ensure compliance with BS5268: Part 5: 1989, it is necessary to refer to Table 4 of that standard. The following notes are intended as guidance on an understanding of Table 4. Further guidance is given in the relevant chapters in this book which deal with typical specifications of specific structural and semi-structural components.

If it is decided that there is a 'risk' but only from attack by House Longhorn Beetles (i.e. no risk of fungal attack) in the roof timbers referred to in the Approved Document to support Regulation 7, and if the specifier wishes to check that the correct schedule is being used, it is necessary to refer to Table 4 of BS5268: Part 5: 1989. From item 1 of this table it can be seen that, for a 'permeable' or 'moderately resistant' softwood, BS5268: Part 5: 1989 could be satisfied by:

- a 20 g/l CCA solution applied by the 'P3' schedule of BS5268: Part 5 (see the later notes on BS5268: Part 5), or
- an organic solvent containing an insecticide, applied by the V/1 double vacuum schedule of BS5268: Part 5, or by a 10 minute immersion.

For a 'resistant' or 'extremely resistant' softwood, BS5268: Part 5: 1989 could be satisfied by:

- a 30 g/l CCA solution applied by the 'P3' schedule of BS5268: Part 5 (see the later notes on BS5268: Part 5), or
- an organic solvent containing an insecticide, applied by the V/1 double vacuum schedule of BS5268: Part 5, or by a 10 minute immersion.

Note that a 20 g/l or 30 g/l solution refers to the number of grammes per litre (i.e. a 2% or 3% solution respectively).

If it is decided that there is a 'risk', but only from attack by common Furniture beetles (i.e. no risk of fungal decay), in, for example, rafters, purlins or ceiling joists in a pitched roof ('dry'), for 'permeable' or 'moderately resistant' softwoods, BS5268: Part 5: 1989 (Item 2(a) of Table 4) could be satisfied by:

- a 20 g/l CCA solution applied by the 'P2' schedule of BS5268: Part 5, or
- an organic solvent containing an insecticide, applied by the V/1 double vacuum schedule of BS5268: Part 5, or by a 3 minute immersion.

For 'resistant' or 'extremely resistant' softwoods, BS5268: Part 5 (Table 4) could be satisfied by:

- a 30 g/l CCA solution applied by the 'P2' schedule of BS5268: Part 5, or
- an organic solvent containing an insecticide, applied by the V/1 double vacuum schedule of BS5268: Part 5, or by a 10 minute immersion.

These notes and conclusions are used, for example, in the build-up to Table A8.1 for the preservation of timber in traditional 'cut' pitched roofs.

If it is decided that it is necessary to preserve against fungal attack as well as House Longhorn Beetles or common Furniture beetles, see the notes below under 'Treatment of Timber in Buildings Against Fungal and Insect Attack'.

When deciding, for example, on the acceptable treatments by an organic solvent for wallplates of 'resistant' or 'extremely resistant' species in a pitched roof ('dry') where there is considered to be a risk of attack by common Furniture beetles but no risk of fungal attack, it can be difficult on a first reading to determine if the intention of Table 4 of BS5268: Part 5: 1989 is to permit a 10 minute immersion. A check with the drafting committee confirms that, provided that there is not considered to be a risk of fungal attack, a 10 minute immersion is acceptable. (Also see Table A5.1 on the preservation of timber for semi-structural uses.)

TREATMENT OF TIMBER IN BUILDINGS AGAINST FUNGAL AND INSECT ATTACK

As stated above, it is not normally necessary for a specifier to go to the extent of specifying the actual preservation schedules which are permitted to be used. However, if a check has to be made to ensure compliance with BS5268: Part 5: 1989, it is necessary to refer to Table 4 of BS5268: Part 5: 1989. The following notes are intended as further guidance on an understanding of Table 4.

If it is decided that there is a risk of fungal attack and insect attack to timber in buildings and there is a need to determine the actual preservation schedules which are permitted to satisfy BS5268: Part 5: 1989, it is necessary to interpret Table 4 of BS5268: Part 5: 1989 to choose a schedule or schedules for the more serious risk combination. To do so, it is necessary to take note of the vertical column which gives the 'risk category' in Table 4 (see Tables B5.4 and B5.5) for which schedules are listed horizontally in Table 4, and also the 'Remarks' column in Table 4 of BS5268: Part 5: 1989 which may refer the specifier to another horizontal line of schedules when there is an additional risk to take into account (i.e. fungal attack). Reference should then be made to Tables 7, 8 and 9 of BS5268: Part 5: 1989 to select the more onerous schedule. (Note that, in the case of the CCA schedules P2, P3 etc., those with a higher number are not always the more onerous.)

To demonstrate, take the example of a 'resistant' species such as European Whitewood used for the rafters, purlins, ceiling joists and ceiling binders in a pitched roof void ('dry') where it is considered that there is a risk of attack by House Longhorn Beetles and a risk of fungal attack because of a local hazard. (Note that tiling battens fixed outside the roof void are not considered to be at risk from House Longhorn Beetles. See Chapter A14 for preservation of tiling battens.)

To give adequate protection against House Longhorn Beetles, horizontal Item 1 of Table 4 lists, for R or ER species, either:

> a 30 g/l CCA Schedule P3, organic solvent Schedule V/1 or a 10 minute immersion.

However, the 'Remarks' column refers the specifier to horizontal Item 2 for cases where there is a fungal decay risk in a pitched roof ('dry').

When one does refer to Item 2 of Table 4, for rafters, purlins, ceiling joists and binders, it can be seen that the only difference in the requirements is that the 30 g/l P3 schedule could be replaced by the shorter 30 g/l P2 schedule. Obviously, to protect from House Longhorn Beetle, if one chooses a CCA schedule, the P3 schedule would be necessary.

These notes and conclusions are used in the build-up to Table A8.1 on the preservation of timber in traditional dry 'cut' pitched roofs.

To preserve a 'resistant' wallplate against fungal attack as well as against House Longhorn Beetles, the same CCA schedule which is necessary against House Longhorn Beetles would be acceptable but, by organic solvent, the longer V/3 or V/4 schedules would be necessary, and immersion is not permitted. (See Items 1 and 2(c).)

Further examples of the use of Table 4 of BS5268: Part 5: 1989 are given in Chapter A7 (Table A7.1) for trussed rafter roofs, Chapter A8 (Table A8.1) for traditional 'cut' roofs, Chapter A9 (Table A9.1) for floor and flat roof joists, Chapter A5 (Table A5.1) for wallplates, Chapter A10 (Table A10.1) for timber stud walls, and Chapter A14 (Table A14.2) for tiling battens.

TYPES OF PRESERVATIVES

As stated earlier in this chapter under 'Amenability to Preservative Treatment', the two principal types of preservatives used in the UK for new building components are:

- The water borne salt types of which by far the most common in the UK are the formulations based on solutions of copper sulphate, sodium dichromate and arsenic pentoxide covered by BS4072: Part

1: 1987 *Wood preservation by means of copper/chromium/arsenic compositions. Specification for preservatives* (usually referred to as copper/chromium/arsenic or simply CCA). They are applied by pressure (sometimes referred to as vacuum/pressure) in a pressure tank. The methods of preserving are covered in BS4072: Part 2: 1987 *Wood preservation by means of copper/chromium/arsenic compositions.*

- Organic solvent based types. There are several different formulations, generally covered by BS5707: Part 1: 1979 *Solutions of wood preservatives in organic solvents. Specification for solutions for general purpose applications, including timber which is to be painted.* The solutions consist of one or more organic fungicides in an organic solvent such as a white spirit or a more refined solvent. Common fungicides are pentachlorophenol, tributyl tin oxide, zinc naphthenate, and pentachlorophenyl laurate. There are others.

In addition, an insecticide should be added if resistance to insect attack is required. It is quite normal to apply organic solvents by a double vacuum process in a pressure tank although some species can be treated for certain applications by immersion in an organic solvent solution. BS5707: Part 3: 1980 *Solutions of wood preservatives in organic solvents. Methods of treatment*, covers the methods of application.

A tar oil type of preservative (e.g. creosote) can be used but such preservatives have a strong smell and can contaminate and are mentioned in this book only as relevant to the preservation of fencing. The associated British Standards are BS144: 1973 *Specification for coal tar creosote for the preservation of timber*, BS913: 1973 *Specification for wood preservation by means of pressure creosoting*, and BS3051: 1972 *Specification for coal tar creosote for wood preservation (other than creosotes to BS144)*. It is the intention that they will be combined in a new version of BS144.

Timber preserved by a diffusion process (e.g. the boron process) can be used for several applications, but the diffusion process is not covered in this book. It must be carried out on timber before it is dried. Most sawmills have discontinued supplies of boron treated timber although research is continuing on advanced methods of application.

Several preservatives are formulated and designed for remedial work. They are not covered in this book.

PRESERVATION CYCLES FOR VARIOUS RISK CATEGORIES

Rather than extend this chapter to virtually repeat BS5589: 1989 and BS5268: Part 5: 1989, the extent of preservation required for the various species covered by this book when the risk categories make preservation 'essential', 'desirable', 'optional' or 'unnecessary' (see Tables B5.4 and B5.5) is covered in the chapters on specifying individual components (i.e. Chapters A1, A2 etc.). A few examples have been given earlier in this chapter under the sub-titles 'Treatment against insect attack' and 'Treatment of timber in buildings against fungal and insect attack'.

GENERAL POINTS ON PRESERVING

If a water borne process is used, the timber should have been dried to about 28% moisture content (see BS4072: Part 2) or less before it is treated. It is extremely important for a specifier to realize that the use of a water borne process increases the moisture content of the timber very considerably, causes an increase in the cross-section, and will raise surface grain. Therefore timber for building usually has to be re-dried before being used and the specifier must allow time and cost for this to be carried out. With some thin components (e.g. small battens) it is usually acceptable to air dry the timber before use, but for certain uses it will be necessary to kiln the timber after preservation.

CCA salts take about seven days to 'fix' in the timber and the timber should not be used before this period is over. The preservative salts are chemically 'fixed' in the timber after treatment. However, soluble salts (e.g. sodium sulphate) which form as a by-product of the reactions between CCA and timber may appear as a white deposit on the surface of the timber.

Specifiers should appreciate that, if a CCA treatment is used on a moulded section, there is a risk of distortion and raising of surface grain. Therefore, if a smooth surface is required, the final machining may have to be carried out after preservation and re-drying has taken place. If, however, a smooth surface is not vital, as for example with cladding boards, the preservation may be carried out with a CCA process after moulding. Moulding after preservation removes part of the preserved timber which defeats part of the object of preserving, and leaves the processor with waste which must be disposed with great care. In the UK, it is normal to use an organic solvent to preserve joinery and moulded sections, and even to use an organic solvent to preserve some construction timber sections where the drying time and surface quality after preservation are important.

It is reported that timber can be preserved by an organic solvent at a moisture content as high as 28%. However, bearing in mind the type of component normally preserved by an organic solvent preservative, it is usual to preserve at moisture contents of 22% or less. Organic solvents do not increase the moisture content of the timber nor affect the dimensions/profile of the sections, nor do they affect the surface to a noticeable extent. For a few years it was thought to be advantageous, when preserving exterior joinery with an organic solvent, to use one containing a water repellant (providing that it was compatible with the finish), but their use is being reconsidered.

If timber treated with an organic solvent is to be painted, it is essential to limit the amount of free solvent left in the timber after preservation. This point is particularly important if the timber is to be factory finished soon after preservation.

As far as possible, all cutting, notching etc. must be carried out before the timber is preserved. If cross-cutting after treatment is unavoidable, the cut ends should be given a thorough application of a suitable preservative. Surfaces exposed by longitudinal conversion cannot be treated satisfactorily by brush application.

An organic solvent preservative does not give protection against soft rot. Therefore, where there is a risk of soft rot (not to be confused with wet rot), as in the case of timber in direct contact with concrete, a water borne CCA preservative must be used rather than an organic solvent.

COMPATIBILITY

The preservative, including any additives, must be compatible with the fixings, adhesive, paint or other finish, weatherseals etc. Cases have been reported of solvent in preservatives reacting with certain weatherseals, and of some water repellants causing problems during application of a water based finish.

Once timber is dried after being treated with a CCA preservative, it is generally not corrosive to metals in normal dry building applications but, if timber is wet or becomes wet, the preservative may increase the rate of corrosion of some metals. Advice on this point should be sought from the manufacturer of individual preservatives. Uncoated aluminium should not be used in contact with preservatives containing copper.

FLAME RETARDANT

Special formulations are available which will raise the 'surface spread of flame' classification of timber to Class 1 when tested according to BS476: Part 7: 1987, or achieve a Building Regulation rating of Class O, which requires a fire propagation index (I) not exceeding 12 and a sub-index (i) not exceeding 6, when tested in accordance with BS476: Part 6: 1981. Such formulations do not improve the 'fire resistance' (charring rate) of timber to any significant extent. Galvanized fasteners should not be used in timber which has been treated with a flame retardant unless the formulation is a non-salt type.

BLUE STAIN/SAP STAIN/YARD BLUE/DISCOLOURED SAPWOOD AND ANTI-STAIN

The sapwood of certain species (e.g. European Redwood) is susceptible to discoloration of the sapwood (i.e. 'sap stain', 'blue stain' or 'yard blue') which may occur if newly sawn timber is left too long before being dried or if, at a later stage, the moisture content of timber is left too long at a moisture content in excess of about 20/22%. Sap stain is not a structural defect and will not spread whilst the timber is at less than about 20/22% moisture content.

Some sawmills treat the sawn timber with an anti-stain chemical although it is a practice which is becoming less popular and less necessary. This is covered in the chapters in this book on supply from various countries (i.e. Chapters C1, C2 etc).

BS4978: 1988 on softwood grades for structural use, permits sap stain in structural timber and, in general, BS1186: Part 1: 1986 on the quality of timber in joinery, permits discoloured sapwood in joinery if an opaque paint is to be used.

THE EFFECT OF PRESERVATIVES ON MOISTURE CONTENT READINGS

The readings given by a moisture meter are affected by CCA preservatives and by flame retardants, therefore the moisture content of such treated timber can be measured only by destructive methods unless the manufacturer of the meter can give a correction factor. Organic solvent treatments do not affect moisture meter readings to any significant extent (with the possible exception of those containing copper naphthenate).

TREATING MIXED SPECIES

When treating mixed species, the normal rule is to use a treatment schedule designed for the species most resistant to preservation. However, this has led to over-absorption in certain cases (e.g. Canadian Spruce-Pine-Fir studs), and NHBC and BRE agreed modified treatment schedules for the BWPA Commodity Specification C9. These schedules are specified in NHBC Practice Note 5 for studs in external walls of timber framed housing, and in BS5268: Part 5: 1989.

EXTERIOR WOOD STAINS OR DECORATIVE STAINS

Finishes given the title 'preservative stains' should not be confused with preservatives. A stain gives virtually no added preservation to the timber, and the term 'exterior wood stain' or 'decorative stain' is to be preferred even if the stain contains a fungicide. The fungicide is intended to prevent mould growth on the surface or in the finish, not to preserve the timber. There are, however, developments of coloured preservatives which act as a decorative stain which may justify the term 'preservative stain'.

SURFACE PHOTO-DEGRADATION

All species of timber left exposed externally (or given a clear finish) will be subject to surface photo-degradation ('greying').

AREAS OF HIGH-POROSITY

In several countries with large sawmilling industries, it has been traditional to store logs in water prior to sawing. The use of decorative stains in recent years has highlighted the fact that biological action in some water can lead to patches or areas of high-porosity on the surface of timber sawn from 'ponded' logs. This is particularly the case with some species (e.g. European Redwood) if logs are stored de-barked in water for several weeks during warm weather. Although this high-porosity (sometimes referred to as 'over-porosity') does not constitute a breakdown of the timber, many sawmills, for example in Sweden and Finland, now use land storage for logs intended for joinery timber. When a stain is applied, areas of high-porosity are highlighted by absorption of far more colour pigment than the surrounding timber and this can lead to an unacceptable appearance.

SAFETY

Preservatives for use in the UK must be cleared through the government's Control of Pesticides Regulations, formerly the Pesticides Safety Precaution Scheme (PSPS). These Regulations came into force in 1986 and made it an offence to sell, supply, store or use a pesticide or wood preservative which has not been approved under the Regulations. One of the conditions is that CCA treated wood must be held for 48 hours after treatment or until dry, before despatch or erection. Obviously preservatives are toxic and must be treated with care during storage and application. Some are flammable until the solvent has evaporated, but thereafter the flammability of the timber is not increased compared with untreated timber. Care must be taken not to taint foodstuffs.

If moulding, cutting etc. is carried out after preservation, great care must be taken to ensure safe disposal of the waste. It should not be used, for example, for animal litter. Mandatory requirements for the safe disposal of wood preservative and treated wood waste are governed by The Control of Pollution Act 1974. For further information regarding recommended methods of disposal, see DOE Waste Management Paper No. 16, *Wood preserving wastes.*

PUBLICATIONS RELATED TO PRESERVATION

BS5268: Part 5: 1989

BS5268: Part 5: 1989 *Structural use of timber. Code of practice for the preservative treatment of structural timber*, is part of the Code of Practice for the structural use of timber. The 1989 version departs from previous editions in that recommended treatment schedules (cycles) and, for CCA treatments, the recommended preservative solution concentrations are given for various treatability and risk category combinations, rather than the previous required retention levels or required penetration depths of preservatives.

The recommended treatments detailed in Table 4 of BS5268: Part 5: 1989 for 'Timber in buildings', assume 'normal risk' and a desired

Table B5.4 Risk categories related to fungal decay in BS5268: Part 5: 1989 for structural timber

Risk category	Risk of fungal decay	Recommended action by BS5268: Part 5
1	Where conditions of use involve negligible risk of fungal attack	Unnecessary to preserve
2	Where there is a low risk of fungal decay	Optional to preserve
3	Where experience has shown that there is a high risk of fungal decay	Desirable to preserve
4	Where timbers are exposed to a continually hazardous environment leading to an unacceptable risk of fungal decay	Essential to preserve
M	Where timbers may be exposed to attack by marine borers	Essential to preserve

The point is made in BS5268: Part 5: 1989 that these risk categories can be used, where appropriate, when considering the risk of insect attack, by substituting the words 'insect attack' for 'fungal decay'. For example, in an area where House Longhorn Beetles are active, softwood in a roof void would be in Risk Category 4. In most other cases of structural timber in a 'dry' covered building, the risk of significant insect attack leading to structural failure is likely to be very low.

Table B5.5 Risk categories related to safety and economic factors in BS5268: Part 5: 1989 for structural timber

Risk category	Safety and economic factors to consider	Recommended action in BS5268: Part 5
A	Negligible safety or economic risk	Unnecessary to preserve
B	Where remedial action or replacement is simple and preservation may be regarded as an insurance against cost of repairs	Optional to preserve
C	Where remedial action or replacement would be difficult and expensive	Desirable to preserve
D	Where collapse of structures would constitute a serious danger to persons or property	Essential to preserve

service life of 60 years. The recommended treatments detailed in Table 5 of BS5268: Part 5: 1989 for 'Timber used for structural purposes other than buildings', are intended to give at least the service lives also detailed in Table 5. Although a component may be preserved for a desired service life performance category, there is no reason to suppose that, at the end of that notional period, it will not continue to perform adequately.

Risk categories 1, 2, 3 and 4 (plus M for 'Marine') are defined related to the risk of fungal decay. These risk categories are described in Table B5.4. The point is made in BS5268: Part 5: 1989 that these categories can be used, where appropriate, when considering the risk of insect attack, by substituting the words 'insect attack' for 'fungal decay'.

Another four risk Categories A, B, C and D are defined related to structural safety and economic factors. These risk categories are described in Table B5.5.

The wording of the risk categories does permit the specifier an element of discretion but does leave an element of doubt. However, once a decision has been taken to preserve, Tables 4 and 5 of BS5268: Part 5: 1989 clarify the treatment schedules which are acceptable, and Tables 7, 8 and 9 detail the schedules. These tables are rather detailed and are not reproduced in this book.

When referring to Table 4 of BS5268: Part 5: 1989, care should be taken to study the notes given in the vertical column on the 'Risk category' for which the schedules are detailed, and also the 'Remarks' column, particularly if the intention is to preserve for both fungal risk and risk of insect attack. It may be necessary to select the more onerous of two treatment requirements. (See the examples given earlier in this chapter under the sub-titles 'Treatment against insect attack' and 'Treatment of timber in buildings against fungal and insect attack'.)

The schedules for CCA treatments are given the Codings P1, P2, P3 or P4. The choice of which to use depends on the risk category and the size of timber. The double vacuum schedules for organic solvent treatments are given the Coding V/1, V/2, V/3, or V/4. The choice of which to use depends on the risk category (although two alternative acceptable schedules are given for V/3 and two for V/4). Immersion treatments in organic solvents are given the Coding M3, M10 and M60 depending on the risk category. (In thc casc of immersion treatments, the figure represents the time in minutes of the immersion.)

When a CCA process is to be used, Tables 4 and 5 of BS5268: Part 5: 1989 give details of the solution which should be used for various components and timbers of different amenability to treatment groups. These used to be referred to as 2% or 3% solutions, but are now given as the number of grammes per litre (i.e. 15 g/l, 20 g/l or 30 g/l).

If a specification clarifies the use of the timber component and the 'risk' or 'risks' (if any) to which it is thought it will be subjected (i.e. House Longhorn Beetles, other insect attack such as from common Furniture beetles, and/or fungal attack), and states that preservation is required, it is possible for the part of the specification which deals with preservation simply to state 'to BS5268: Part 5: 1989'. That, however, might give the supplier or manufacturer rather more license than the specifier is prepared to accept, and it is usually preferable for the specifier to state or agree if a CCA schedule and/or an organic solvent double vacuum or immersion schedule is required. (See the examples given in Chapters A7, A8 etc.)

The V/4 schedules are new to the preservation industry and are shown in BS5268: Part 5: 1989 as preferred alternatives to the corresponding V/3 schedules. Before specifying or insisting on one of the V/4 schedules being used rather than the corresponding V/3 schedule, it would be prudent to consult the companies which are providing the preservative and are involved with preserving timber. The general feedback from industry is that the V/4 schedules will lead to problems due to over-absorption of solvent.

BS5589: 1989

BS5589: 1989 *Code of practice for preservation of timber*, is complementary to BS5268: Part 5 in that it aims to cover non-structural timber in buildings and other uses such as fencing (although it does duplicate one part of BS5268: Part 5: 1989 in that it covers preservation of timber in the exterior stud walls of timber framed houses). It gives tables of use conditions (e.g. external woodwork in buildings above the damp-proof course), leaves it to the specifier or user to decide on a desired service life performance category, and then gives suitable treatment schedules or immersion periods for various species. Thus, if a specifier quotes BS5589: 1989 and wishes to be precise, it is necessary also to quote the desired service life performance category (other than when specifying preservation of timber for use in termite infested areas, and for timber in the exterior stud walls of timber framed houses for which no optional performance categories are given). The optional performance categories in the tables in BS5589: 1989 are:

Table 5. Preservative treatment for external woodwork in buildings, above the damp-proof course—30 or 60 years desired service life (i.e. Performance Category B or A respectively).

Table 7. Preservative treatments for timber in contact with the ground, soil or manure; also those likely to become and remain wet—20 or 50 years desired service life (i.e. Performance Category B or A respectively).

Table 8. Preservative treatments for timber subject to intermittent wetting or damp conditions—20 or 50 years desired service life (i.e. Performance Category B or A respectively).

Table 10. Preservative treatments for timber for use in contact with water—30 years in fresh water, 15 years in sea water.

Table 11. Preservative treatments for European Redwood timber for use as packing in cooling towers—15 or 30 years desired service life (i.e. Performance Category B or A respectively).

Table 14. Preservative treatments for fencing timber in Performance Category A (desired service life 40 years).

Table 15. Preservative treatments for fencing timber in Performance Category B (desired service life 20 years).

Although a component may be preserved for a desired service life performance category, there is no reason to suppose that, at the end of that notional period, it will not continue to perform adequately.

BS5589: 1989 is discussed further in the relevant chapters on components in this book (e.g. Chapters A1, A2 etc.).

BS1186: Part 1: 1986

Although BS1186: Part 1: 1986 *Timber for and workmanship in joinery. Specification for timber* does not cover preservation in detail, it is important in considerations of durability and preservation in that Appendix B lists six species of softwood and states whether they are:

S Suitable (without being preserved),
SP Suitable if preserved in accordance with BS5589, or
X Unsuitable (even if preserved).

In Appendix B of BS1186: Part 1, only Western Red Cedar, Douglas Fir, Western Hemlock, Parana Pine, European Redwood and European Whitewood are listed. In general, none of these species are considered suitable for door sills or thresholds (unless covered). Douglas Fir and Western Red Cedar are rated *S* for exterior joinery (unless sapwood is present), and European Redwood, European Whitewood and Western Hemlock are rated *SP* for exterior joinery. Parana Pine is rated as *X* for exterior joinery.

A warning note is given about resin exudation under dark finishes, and about surface photo-degradation of exterior surfaces if a non-opaque finish is applied.

BRE Technical Note No. 24

With the updating of British Standards and Codes dealing with preservation of joinery, BRE Technical Note No. 24 *Preservative treatments for external softwood joinery timber* has been withdrawn. It used to be regarded as the authoritive document for the preservation of certain external joinery items such as softwood windows although, in turn, it referred to Section 2 of BS5589: 1978. Although it did not go into such detail on treatment schedules as BS5589: 1978, it contained useful reading for specifiers in that Appendix B gave a list of available commercial preservatives which were approved for use in treating external joinery by water borne or organic solvent methods. Although it has been withdrawn, it is mentioned here in case a specifier is involved with an extension to a building which has timber preserved in accordance with Technical Note No. 24.

Commodity Specifications of the British Wood Preserving Association

Several useful publications are available from the BWPA. Amongst these there are nine 'Commodity Specifications' for various softwood applications as listed below:

C1 (1986) Preservative treatment of timber to be used as packing in cooling towers.

C2 (1986) Preservative treatment of timber for use permanently or intermittently in contact with sea or fresh water.

C3 (1986) Preservative treatment of fencing timber.

C4 (1986) Preservative treatment of agricultural and horticultural timbers.

C5 (1986) Preservative treatment on non loadbearing external softwood joinery and external fittings (excluding cladding) not in ground contact.

C6 (1986) Preservative treatment of external timber cladding.

C7 (1986) Preservative treatment for timber for use in buildings in termite infested areas.

C8 (1986) Preservative treatments for constructional timbers (excluding walls of timber frame houses).

C9 (1986) Preservative treatments for timber frame housing.

BWPA Commodity Specifications are offered to BSI for consideration in the drafting of British Standards. Therefore, if the date of the relevant BS is later than the date of the BWPA Commodity Specification, it is probably better to consider the BS as reflecting the latest thinking.

Building Regulations

The Building Regulations 1985 for England and Wales are not as precise as the previous Regulations in stating when preservation is or is not required to satisfy the Regulations. In general, it can be said that Clause 1.2 on 'Ways of Establishing Fitness of Materials' and Clauses 1.3, 1.4 and 1.5 'Short-lived Materials' of the Approved Document to support Regulation 7 will be satisfied if the relevant British Standards and Codes of Practice are satisfied. Clause 1.4 which covers short-lived materials which are 'readily accessible for inspection and maintenance and replacement' may be read as giving an authority not to preserve in certain cases where the relevant code implies the need for preservation (e.g. exterior timber cladding on a bungalow), or where NHBC call for preservation.

As stated earlier in this chapter on the 'Risk and Avoidance of Insect Attack', The Building Regulations 1985 for England and Wales, plus the Approved Document to support Regulation 7 combine to require adequate treatment with a suitable preservative of softwoods in roof voids in the areas of the Home Counties where House Longhorn Beetles are a risk.

The Building Standards (Scotland) Regulations 1981 have two deemed-to-satisfy provisions which are relevant to the preservation of timber. In the case of timber weatherboarding fixed direct to studs, any Redwood or Whitewood from Northern Europe, Western Hemlock, Californian Redwood, Eastern Canadian Spruce, Western Canadian Spruce, Douglas Fir, or British-grown Scots Pine can be used if it is preserved in accordance with BS5589. (See Schedule 13.G9 (6) and

Schedule 14. Part 1. Item 8 of The Regulations). The Regulations do not require Western Red Cedar to be preserved. Where weatherboarding is fixed in a position where it is readily accessible for inspection and maintenance or renewal, it can be claimed that Regulation B2(ii) makes preservation unnecessary, particularly if the weatherboarding is fixed to battens.

The second deemed-to-satisfy provision is Schedule 13.G9 (23) in which Canadian Western Red Cedar shingles are permitted to be used on a roof slope of not less than 14° if preserved in accordance with BS4072.

In general, The Building Regulations (Northern Ireland) 1977 will be satisfied if the relevant British Standards and Codes of Practice are satisfied. The Table to Regulation B3 and Tables 2 and 3 call for the softwoods which are covered in this book, except Western Red Cedar, to be preserved if used as exterior timber boarding (cladding).

The Registered House-Builder's Handbook of NHBC

In their Handbook and in their Practice Note 5 for timber framed housing, NHBC call for preservation as required to satisfy Building Regulations and, in addition, for preservation of the following items:

	CCA	Double vacuum organic solvent
Lintels in brick or blockwork external walls	+	–
Battens as fixing for claddings	+	+
Any embedded timber	+	–
Joists in flat roofs	+	+
End of joists with ends built into solid (non-cavity) walls	+	+
Door frames (for external doors)	+	+
Windows	+	+
Surrounds to metal windows	+	+
External doors other than flush doors	+	+
External timber features other than fencing	+	+

+ = permitted; – = not permitted.

OTHER RELEVANT BRITISH STANDARDS DEALING WITH PRESERVATION

BS1282: 1975 *Guide to the choice, use and application of wood preservatives.*

BS4072: Part 1: 1987 *Wood preservation by means of copper/chromium/arsenic compositions. Specification for preservatives.* (AMD 6200 published in June 1989 permits specifications quoting BS4072: Part 1 to be satisfied by oxide type CCA formulations.)

BS4072: Part 2: 1987 *Wood preservation by means of copper/ chromium/ arsenic compositions.*

BS5707: Part 1: 1979 *Solutions of wood preservatives in organic solvents. Specification for solutions for general purpose applications, including timber that is to be painted.*

BS5707: Part 2: 1979 (1986) *Solutions of wood preservatives in organic solvents. Specification for pentachlorophenol wood preservative solution for use on timber that is not required to be painted.*

BS5707: Part 3: 1980 *Solutions of wood preservatives in organic solvents. Methods of treatment.*

BS144: 1973 *Specification for coal tar creosote for the preservation of timber.* To be rewritten as Part 1 on the specification of preservatives, and Part 2 on methods of application. The intention is that the rewrite will incorporate an update of the previous BS144: 1973, BS913: 1973 *Specification for wood preservation by means of pressure creosoting*, and BS3051: 1972 *Specification for coal tar creosotes for wood preservation (other than creosotes to BS144).*

PERFORMANCE IN FIRE B6

INTRODUCTION

Timber is a combustible material and, despite much research, no commercial way has yet been found to render it incombustible. This fact has often led specifiers to believe that timber is unsuitable as a fire resisting structural material but that is far from the truth.

Although combustible, timber is not easy to ignite in the sizes normally encountered in building. Once ignited, timber burns very slowly, and builds up a protective layer of charcoal on its surface which insulates the remainder of the section from the worst effects of fire. This protective shield, plus the excellent thermal insulation property of timber, combined with a very low coefficient of thermal expansion and properties largely unaffected by elevated temperature, allowed a calculation method to be derived which can be used to ensure that the residual section after charring can perform its structural function even if fully exposed and subjected to fire. Depletion of timber in a fire is slow and predictable, and has been widely researched over many years. It has been established by research, for example, that a softwood section will deplete by around 20 mm in thickness and width in half an hour during a 'standard fire test' when fully exposed.

Armed with facts about the charring rate, and knowledge that the strength of a residual section is relatively unaffected by exposure to heat, it is possible to design a section for fire safety. This design method is now detailed in BS5268: Part 4: Section 4.1: 1978.

Although timber members can be designed for full exposure in a fire situation, it is a much more common condition for the timber to be concealed by other building materials (e.g. plasterboard in a timber stud wall). Where these materials have fire resistance in themselves they may be taken into account in determining the fire resistance of the element as a whole. Until recently, it was necessary for a fire resistance test to BS476: Part 8: 1972 to be carried out as the only means of validating this shielding effect and, of course, the accumulated results of this testing is used as the basis for UK Regulations and other authoritative references. Recently published, however, is a new BS5268: Part 4: Section 4.2: 1989 on a method of calculating fire resistance of timber stud walls and joisted floor constructions, which permits designers to aggregate the various contributions to fire resistance of the individual parts of an element.

Fire resistance is not the only fire safety requirement in modern Building Regulations. Many wall and ceiling surfaces are required to provide resistance to 'surface spread of flame'. The natural surface rating for most softwoods is Class 3 as defined in BS476: Part 7: 1987 and, in many cases, the surface has to be up-rated by the use of chemical treatment when Regulations call for internal exposed surfaces to be of Class 1, or Class 0 as defined in The Building Regulations 1985 for England and Wales.

Other Regulations concerning fire safety which effect timber in use relate to:

- the proximity to boundaries,
- 'unprotected area' rules, and
- cavity-barriers and fire-stops.

The following notes are intended to familiarize the reader with the various aspects of fire terminology and how these relate to current building legislation. For more detailed information, the reader is referred to the various British Standards and Regulations referenced in this chapter, and to the booklet *Performance in Fire*, published by The Swedish Finnish Timber Council from which extracts have been used in the following text by permission of SFTC.

FIRE RESISTANCE

Definition of fire resistance

In recent years, four new Parts to BS476 have been published with the intention that they should replace BS476: Part 8: 1972 on the fire resistance of elements of building construction. The new Parts are:

BS476: Part 20: 1989 on the general principles for determining the fire resistance of elements of construction.

BS476: Part 21: 1987 on the fire resistance of loadbearing elements of construction.

BS476: Part 22: 1987 on the fire resistance of non-loadbearing elements of construction, and

BS476: Part 23: 1987 on the contribution of components to the fire resistance of a structure.

Currently, however, BS476: Part 8: 1972 is still 'cited in legislation' (to quote BSI) and is still current, not having been deleted yet by BSI. In view of that, this chapter and other parts of this book which cover fire resistance of elements of building construction still refer to BS476: Part 8: 1972.

The fire resistance of an element of construction is the period of time for which it is able to meet the criteria laid down in a standard fire test to BS476: Part 8: 1972. In BS476: Part 8: 1972, the fire resistance of an element is assessed in relation to three performance criteria:

'Stability'
'Integrity'
'Insulation'.

- *'Stability'* is the resistance to structural collapse. The element must support the applied load throughout the test and for 24 hours after the removal of the heat source. In the case of a beam it must also deflect no more than 1/30 of the clear span.
- *'Integrity'* is the resistance of an element to perforation, which would permit flame or hot gases to pass through.
- *'Insulation'* is the resistance to passage of heat which would cause the unheated face to rise to an unacceptable temperature.

Not all elements or components need satisfy all three criteria. For example, a fully exposed beam or column has only to satisfy the 'stability' criteria. Conversely, a wall designed to contain a fire is required to satisfy all three criteria. In some cases, however, Building Regulations permit a relaxation of one or more criteria. For example, the Regulations for an intermediate floor in a two-storey dwelling require what is referred to as a 'modified half hour' floor because the criteria for 'integrity' and 'insulation' are relaxed to 15 minutes whilst the floor must satisfy the 'stability' criteria for a full 30 minutes. Tables in Building Regulations stipulate what periods are required for each of the criteria in relation to various elements of structure. For example a 30/15/15 requirement is for 30 minutes 'stability', 15 minutes 'integrity' and 15 minutes 'insulation'.

Calculation of fire resistance of solid timber members

There are three alternative methods of satisfying the fire resistance requirements of exposed solid timber members. They are:

- an actual fire test;
- an assessment by a fire testing authority;
- calculations based on the recommendations of BS5268: Part 4: Section 4.1: 1978.

BS5268: Part 4: Section 4.1: 1978 defines a method of calculating fire resistance based on well-established depletion rates for various species exposed to fire. This is sometimes called the 'sacrificial timber' method.

Research has shown that, when a timber member over a certain size is fully exposed to a 'standard fire test', it chars in a reasonably predictable way as indicated in Fig. B6.1. With the exception of

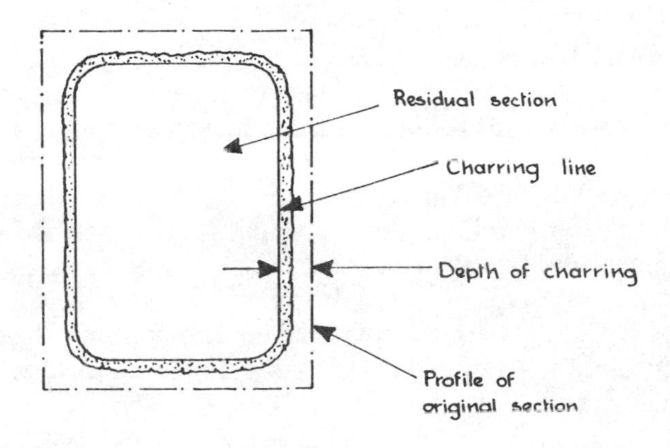

Fig. B6.1 Charring profile of a timber section

Western Red Cedar, the softwoods covered by this Book have a 'notional' charring rate of 20 mm in 30 minutes or 40 mm in 60 minutes. The corresponding figures for Western Red Cedar are 25 mm and 50 mm respectively. Interpolation for different times is permitted as is extrapolation between 15 and 90 minutes. The charring rates may be taken as being uniform over the charring time.

As timber burns, it builds-up a layer of protective charcoal. Fire tests have shown that, with an air temperature of about 900°C at the outside surface of a member, the temperature immediately under the charcoal is less than 200°C. Because of the protective layer and the very low coefficient of expansion of timber, the unit strength of the uncharred timber can be considered to have been unaffected by the fire.

These facts have been established with such certainty that they have been incorporated in BS5268: Part 4: Section 4.1: 1978, and enable a designer to calculate if the residual strength of an exposed timber member is adequate to resist the design load, and to match the deflection criteria throughout the required fire resistance period.

Fire resistance of composite structural elements

When, for instance, a wall or floor construction is required to provide fire resistance, the combination of materials makes a simple calculation method more difficult. Again there are three alternative methods of satisfying the requirements. They are described below.

A fire test incorporating the actual combination of materials to be used is always possible but, before proceeding, it is advisable to check with the Fire Research Station, TRADA etc., and indeed, check the 'Approved' constructions within the Building Regulations themselves, to ensure that the test has not been carried out already.

As an alternative, provided the construction is sufficiently similar to one already tested, a fire testing authority may give an assessment of its fire resistance.

In addition, BS5268: Part 4: Section 4.2: 1989, provides designers with a method of checking most wall and floor constructions based upon aggregating the various contributions of the component parts. For example, in a typical external timber framed wall, the standard permits contributions from the plasterboard inner lining, the studs, the thermal insulation, the sheathing plywood, and even the brick exterior skin if there is one.

SURFACE SPREAD OF FLAME

BS476: Part 7: 1987 on the surface spread of flame of products, gives classifications for the rate of spread of flame over the surface of materials, and a standard test by which materials are judged. The performance indices are:

- Class 1 Surfaces of very low flame spread;
- Class 2 Surfaces of low flame spread;
- Class 3 Surfaces of medium flame spread;
- Class 4 Surfaces of rapid flame spread.

In addition, UK Building Regulations introduce a Class 0 surface category which relates to materials which have passed the test for Class 1 of BS476: Part 7: 1978 and have been subjected to the fire propagation test of BS476: Part 6: 1981, and have achieved two index ratings as required by the Building Regulations. (Fire propagation is a measure of the ignition characteristics, the amount and rate of heat release, and the thermal properties of a product related to their effect on the rate of fire growth.)

With the exception of Western Red Cedar, which has a Class 4 rating due to its low density, the softwoods covered by this book have a density in excess of 400 kg/m^3 and fall naturally into surface spread of flame Class 3, and can be up-rated by special chemical treatments to Class 1 or Class 0. Such treatments are called 'flame retardants'. They have virtually no effect on the charring rate and therefore are not 'fire retardants'.

Surface spread of flame requirements are laid down in UK Building Regulations. They are undoubtedly relevant for certain areas (e.g. corridors) and certain buildings, but there is considerable doubt about their real relevance for safety in many other areas, such as rooms in domestic buildings.

If a flame retardant treatment is used, the compatibility with adhesives, finishes etc, must be checked. Some adhesives cannot be used on timber which has been pre-treated with flame retardants. Some flame retardants are reported as reducing the strength of timber by about 15%. The specifier should check with the manufacturer in each case, and also check the suitability before specifying a flame retardant for an exterior component or for an area of very high humidity. Few are suitable.

COMBUSTIBILITY

BS476: Part 4: 1970 (1984) defines a test to determine if a material is 'combustible' or 'non-combustible'. Timber is defined by test as being combustible and there is no commercial way for building timber to be made non-combustible.

Approved Document B of The Building Regulations 1985 for England and Wales introduces a new definition of 'limited combustibility'. This includes all non-combustible materials and such other materials having a density over 300 kg/m^3 which meet certain requirements when tested in accordance with BS476: Part 11: 1982 which covers a method of assessing the heat emission from building materials. Materials of 'limited combustibility' include such materials as plasterboard which, due to the paper faces, cannot strictly be defined as being non-combustible. Timber cannot be modified to be of 'limited combustibility'.

IGNITABILITY

BS476: Part 5: 1979 defined a test for 'the determination of the ignitability characteristics of the exposed surfaces of essentially flat, rigid or semi-rigid building materials or composites, when tested in the vertical position'. On the basis of this test, surfaces used to be defined as being: 'easily ignitable' or 'not easily ignitable'. These terms are no longer used in the 1979 edition of BS476: Part 5 but the test classifications (i.e. 'X' and 'P') generally signify similar levels of ignitability (i.e. Classification 'P' is similar to the previous 'not easily ignitable').

Timber in the sizes normally used for building purposes used to be defined by test as being 'not easily ignitable'. A test carried out in 1988 by the The Warrington Fire Research Centre on behalf of The Swedish Finnish Timber Council confirmed that Nordic softwood having a density of at least 470 kg/m^3 at 12% moisture content has an ignitability classification of 'P'.

Tests show that, without a source of flame or prolonged pre-heating, timber will not ignite spontaneously until a temperature of around 450–500°C is reached (i.e. the temperature close to that at which steelwork and aluminium lose much of their strength). If, however, timber which has a structural or semi-structural function is to be used at a temperature in excess of about 60°C for a considerable period of time, expert advice should be sought on the permissible stresses to use in the design.

CAVITY-BARRIERS AND FIRE-STOPS

To satisfy various UK Regulations, where either of the interior surfaces of a cavity or void is combustible, cavities or voids in an 'element of structure' must be subdivided to limit the spread of flame and smoke within the cavity. In Building Regulations, the subdivision is by a device called a 'cavity-barrier'. Timber at least 38 mm thick is classified as a suitable cavity-barrier and, as such, may be used to close or subdivide a cavity or void.

A 'fire-stop' is a non-combustible filling used to seal an 'imperfection of fit', such as will occur when a pipe passes through a floor or wall, in an element required to have fire resistance.

ELEMENTS OF STRUCTURE

External walls

In satisfying UK Building Regulations, external walls may be constructed of combustible materials unless they are situated closer to a boundary than 1 m or are over 15 m in height. Walls in houses up to three storeys may be constructed of timber when closer than 1 m to a boundary but, in this case, they must have fire resistance from both sides. In The Building Regulations 1985 for England and Wales, a Class 0 exterior cladding is required. The Building Standards (Scotland) Regulations 1981 require a non-combustible cladding.

Irrespective of other requirements, a minimum period of half hour fire resistance is required in the loadbearing parts of an external wall except those of single storey buildings.

Floors other than compartment floors

Intermediate floors in dwellings are expected to provide fire resistance. In the case of a two storey dwelling, a 'stability' of 30 minutes is required although the 'insulation' and 'integrity' criteria are reduced to 15 minutes. This is why such a floor is referred to as a 'modified half hour' floor.

Provided that a floor in a building of more than two storeys but not more than four storeys in height is not a 'compartment floor' (see below), it may be constructed of timber, but requires a full half hour fire resistance (i.e. 30/30/30).

Compartment walls and compartment floors

In UK Building Regulations, a 'compartment wall' or 'compartment floor' is defined as a wall or floor which divides a building into stipulated areas. A 'compartment wall' may include a door, provided that the door has 'integrity' as required by the relevant Building Regulations.

In England, Wales and Northern Ireland, compartment walls and compartment floors which are required to have a fire resistance of 1 hour or more must normally be constructed from non-combustible materials or materials of 'limited combustibility'. The exceptions are compartment walls between flats of up to three storeys and compartment floors between flats of up to four storeys. Compartment walls and compartment floors requiring only half hour fire resistance may be constructed of timber to the rules given in Approved Document B.

In Scotland, compartment walls may be constructed with timber. Compartment floors may be constructed from timber in flats of up to four storeys.

Where combustible material is permitted in a compartment wall, the same basic considerations apply as for a 'separating wall' (see below).

Roofs

There is no requirement for roofs to have fire resistance from inside a building.

There are rules concerning the performance of roof covering

materials, and surface spread of flame requirements on ceilings underneath the roof.

Interior walls (other than compartment walls)

Timber may be used without restriction in interior walls which are not specifically designated as compartment walls. If an interior wall is loadbearing, it is required to have the same fire resistance as the element of structure which it supports.

Surface spread of flame requirements are detailed in Building Regulations.

Separating walls

A 'separating wall' is a wall or part of a wall which is common to adjoining buildings. Unlike a compartment wall it cannot have a door.

All UK Building Regulations have a basic requirement that separating walls should be constructed from non-combustible materials or materials of 'limited combustibility'. However, the Regulations for England, Wales, Scotland and Northern Ireland make exceptions and permit combustible materials to be used in separating walls between dwellings within certain defined sizes.

The primary requirement of a separating wall is that it will remain imperforate and form a complete vertical separation between the buildings separated, including the roof spaces. No combustible material is permitted to be carried through, into or across the end of, or over the top of, any separating wall in such a way as to render ineffective the fire resistance of the wall.

The minimum period of fire resistance for a separating wall is 1 hour. This can be achieved with timber construction by the well-established principle of constructing the wall basically of two separate frames with adequate plasterboard linings on each.

TIMBER EXTERIOR CLADDING

UK Building Regulations limit the use of combustible materials on the outer wall of a building if it is close to a boundary or if it is above certain heights.

The Building Regulations 1985 for England and Wales generally require any exterior wall closer than 1 m to a boundary or over 15 m high to be constructed of non-combustible materials or materials of 'limited combustibility'. It is possible, however, to use timber as external cladding, even if it is closer than 1 m to the boundary, if the surface is up-rated to Class 0 and if the wall has the specified fire resistance from both sides. (Specifiers should check the suitability of any treatment claimed to up-rate timber from Class 3 to Class 0 for exterior use.)

Walls more than 1 m from a boundary are not required to have exterior fire resistance or a Class 0 exterior surface unless more than 15 m high and, even then, the part below 15 m is not required to be Class 0.

In the terms of The Building Regulations 1985 for England and Wales, timber cladding is an 'unprotected area' and hence the distance it must be from a boundary for the purpose of satisfying Regulations is dictated by the area of the cladding plus any window, door or other opening. The rules for calculating acceptability are too extensive to repeat here. (See Appendix J to Approved Document B to The Building Regulations 1985 for England and Wales.)

The Building Standards (Scotland) Regulations 1981 exclude any combustible cladding material from a wall closer than 1 m to a boundary. If 1 m or further from the boundary, timber cladding may be used without limitation on surface Class but subject to the type of 'unprotected area' rules described above. (See D18(6)(a) and Schedule 8 of the Scottish Regulations.)

FIRE DOORS

When a fire door is required to satisfy Building Regulations, the latest thinking, which is likely to be incorporated into all UK Regulations, is that a door or door assembly need satisfy only a test for 'integrity'. The 1985 version of the Regulations for England and Wales takes this into account, although the Regulations for Scotland and those for Northern Ireland still call for tests for 'stability' and 'integrity' to be satisfied. Future Regulations may call for doors to act as a barrier to smoke.

BS476: Part 22: 1987 covers the 'methods for determination of the fire resistance of non-loadbearing elements of construction'. The intention is that this Part of BS476 (and Parts 20, 21 and 23) will eventually supersede Part 8: 1972. However, Part 8 is still cited in legislative documents and therefore has not been deleted by BSI.

Although Part 8 has not been deleted, feedback from industry is that any test of a new design is being carried out to Part 22 (although repeat testing of an existing design may be to Part 8). Also, feedback from industry is that the 'integrity' test to Part 22 is slightly more onerous than the 'integrity' test in Part 8. (Note that Part 22 does not contain a test for 'stability'.)

Table A1 of Approved Document B2/3/4 to support The Building Regulations 1985 for England and Wales gives minimum provisions for fire doors when tested in accordance with BS476: Part 8: 1972. Only the test for 'integrity' has to be satisfied. The requirement is either for 20 minutes, 30 minutes, the period for the wall in which the door is situated, or half the period for the wall in which the door is situated. See Table A1 of B2/3/4 for the actual periods to be satisfied.

For the periods of 'stability' and 'integrity' which must be satisfied to comply with The Building Standards (Scotland) Regulations 1981, see Regulations D6, D8(1)(i), D9(6), D10(3), E2(6) and Table 1 to E2, E10(7), E10(11), E11(4) and E14.

For the periods of 'stability' and 'integrity' which must be satisfied to comply with the Building Regulations (Northern Ireland) 1977, see table to Regulation E1, E9(1)(a)(i), E10(7)(a), E10(7)(b), E11, E13(2)(b) and E18(6)(c)(ii).

Designs of timber fire doors are available for periods of 20, 30 and 60 minutes (see Chapter A2). In view of the differences in various Regulations, when selecting or specifying a fire door, it is prudent to make it clear if the requirement is for 'integrity', or for 'stability' and 'integrity'. Manufacturing members of the British Woodworking Federation (BWF) have adopted a coding of FD20, FD30 and FD60 for a 20, 30 or 60 minute resistance period respectively as referenced in BS5588: Section 1.1: 1984. The figure refers to the period in minutes of 'integrity' which sample doors have passed in tests to BS476: Part 8 (and/or Part 22). (If 'stability' is required, the specifier or purchaser should check if the design will satisfy the requirements.)

FD30 and FD60 doors are likely to require an intumescent strip to

Table B6.1 Colour code of fire door plugs

Colour code	Without an intumescent strip being fitted	With an intumescent strip being fitted
Green core on white background	FD20	
Green core on yellow background	FD30	
Green core on blue background	FD60	
Red core on white background		FD20
Red core on yellow background		FD30
Red core on blue background		FD60
Blue core on white background	FD20	FD30

be positioned in the door or door frame (or door lining), but see Table B6.1.

BWF manufacturing members have agreed a method of identifying fire doors by the insertion of a circular colour-coded plastic plug in the hanging stile of the door, approximately 600 mm from the top or bottom of the door. By means of different colours, the plug is used to denote the fire resistance of the door. Each plug has a 'background' colour and a circular 'core' colour.

Assuming that the door assembly is correctly assembled (using an intumescent strip where required), the colours of the plug indicate that the fire resistance 'integrity' rating achieved by the design of door in a test to Part 22 or Part 8 is as shown in Table B6.1. Note in particular the link between a red core and the requirement for an intumescent strip to be fitted.

OTHER CONSIDERATIONS

Glued sections

If a glued section such as glulam (see Fig. A11.1) is manufactured using an adhesive of type:

resorcinol-formaldehyde,
phenol-formaldehyde,
phenol/resorcinol-formaldehyde,
urea-formaldehyde or
urea-melamine-formaldehyde,

the section may be considered to act in a fire as though it is a solid section.

Metal fasteners

Any metal fastener which is exposed from the commencement of a fire, or which becomes exposed to fire during the duration of a fire, will tend to conduct heat into the timber and lead to local charring around the fastener, reduction in retention, and reduced strength of the fastener. If required to retain its strength during the period of a fire, a fastener must be positioned below the anticipated charring line or be protected in other ways.

If a metal connection is an integral part of the design for overall stability of a structure, the effect of fire on the connection must be considered in checking the overall stability as well as in checking the local strength.

PAINTING B7

INTRODUCTION

These notes on painting are included because paints of a quality unsuitable for exterior use on timber have been applied to windows, cladding etc. in recent years and have led to a great deal of trouble. In many cases this has led to criticism of the timber rather than the paint. Although there may well have been several contributing factors such as the use of poor details, troubles on exterior surfaces can be traced largely to the use of general-purpose paints (rather than exterior-quality paints) since the virtual ban on lead paints. Lead paints had a certain extensibility and a thickness which worked well with timber; when applied to end grain they had good sealing properties; and it is also likely (although not absolutely established) that they had a toxic effect on any mould which tried to form on the surface of the paint.

The majority of paints which were used initially (and until the mid-1980s) to replace lead paints were general-purpose alkyd paints. These have been found, by and large, not to have sufficient extensibility or the other qualities required for use on exterior surfaces of timber, although they may well be suited for use on surfaces inside a building. Where troubles caused by the use of poor quality paints and poor details led subsequently to components (e.g. timber windows) being preserved, this was a case of treating the effect not the cause.

Exterior-quality paints are now readily available in the UK and several of the following notes are intended as guidance on their use. Of necessity the notes are of a general nature because each paint manufacturer uses variations on basic formulations, and binders and pigments of different quality. It can be stated, however, that general-purpose undercoats and general-purpose top coats *should not be used externally on timber*, and any primer should be of high quality, i.e. to BS5082: 1986 or BS5358: 1986 or better.

For the purpose of this chapter, terms such as 'general-purpose' alkyd paints are used which have the meanings as set out below:

- *General-purpose alkyd.* An alkyd paint sold without specific reference to it having been formulated for exterior use on timber.
- *Exterior-quality alkyd.* An alkyd paint specifically formulated for exterior use on timber. (Quite often such paints are referred to as

being 'microporous'. Some do not have the same degree of gloss as a general-purpose alkyd, although some do.)

- *Exterior-quality acrylic emulsion.* A water based opaque paint formulated with a resin selected to make it suitable for exterior use. (Acrylics are more permeable than alkyd paints of the same thickness. They are slightly thermoplastic therefore can be a little 'tacky' in service.)
- *Exterior wood stain.* A finish which is non-opaque to a greater or lesser extent depending on the pigment and thickness. Such a stain should not be referred to as a 'preservative stain' (unless, of course, formulated to be a preservative). Any fungicide in the stain is not intended to preserve the timber, but to prevent disfigurement of the decorated appearance of the exterior surface by mould fungi. The broad term 'decorative stain' is used, but it should be realized that some decorative stains are suitable for external use, whilst some are not. Most stains used in the UK are solvent based alkyds although water based acrylic stains are available.

PERMEABILITY

Particularly where the finish is applied to an exterior surface of timber which may have picked up moisture (as can happen with site painting), or where moisture may enter through joints (as can happen with exterior joinery or cladding), a degree of permeability may be beneficial to the finish. If a non-permeable paint is applied to a moist surface where there may be a build-up of moisture, the paint is likely to lose its adhesion and be lifted off by the moisture. Each paint manufacturer has different formulations but, as a general guide, it is probably sufficient to regard the various types of finish as having the following degree of permeability:

- *General-purpose alkyds*—largely impermeable.
- *Exterior-quality alkyds*—slightly permeable. (They may be referred to as being microporous or micro-vapour permeable.)
- *Exterior-quality acrylic emulsions*—having a permeability between an alkyd paint and a stain finish.
- *Exterior wood stains*—permeable to an extent depending on the film thickness. A 'low-build' stain is very permeable. A 'high-build' stain is less permeable. The thickness of finish at which a stain ceases to be considered low-build is in the order of 20 microns.

The thickness of the finish affects the permeability, therefore it is important that paint or stain is applied to the correct thickness. (One micron is one millionth of a metre.)

PAINTING OF INTERIOR SURFACES

Providing that the timber is dried to an appropriate moisture content before being painted, and that the surfaces are not subject to any particular hazard in service, general-purpose paints, stains or varnishes, depending on choice, are usually suitable for interior surfaces. If it is relevant to use general-purpose primers or undercoats (for example under an opaque finish), they are generally suitable for interior surfaces. Where the component being painted (e.g. a timber window) forms part of the exterior wall of an inhabited building, there are advantages if the paint on the interior surface is largely impermeable to act as a vapour check, or is at least as impermeable as the exterior finish. The timber, particularly if preserved, will be able to cope with reasonable changes in moisture content providing that moisture is not trapped.

If a clear finish is used, the timber will darken, yellow or 'mellow' depending on the amount and type of light. (One possible exception is when European Whitewood is used because this species retains its original colour to a remarkable extent.)

If the timber to be painted is in a place where it is liable to become dirty, the finish should be chosen to be easy to clean. By and large, high-gloss paints and high-gloss varnishes are easier to clean than stains or matt finishes.

If a paint, stain or varnish is to be used in a bathroom or similar area where the relative humidity can be high on occasions, a formulation containing a fungicide is preferred. Also, care should be taken in detailing the timber to eliminate ledges etc. where moisture can lay or enter the timber (particularly the end grain of timber) by capillary attraction.

When timber boarding is fixed to battens, the battens should be arranged to ensure ventilation behind the boarding.

PAINTING OF EXTERIOR SURFACES

The detailing of timber components used externally is very important to minimize hazards to the paint. In particular, details must be designed to shed moisture quickly, and to be ventilated. The following notes are interdependent.

Primer on exterior surfaces

If a separate primer is used, it must be of adequate quality or the paint system will have short life no matter what other care is taken. If a solvent based primer is used it must be of a quality at least in accordance with BS5358: 1986. If an acrylic primer is used it must be of a quality at least in accordance with BS5082: 1986. Such solvent based or water borne acrylic primers are required to be able to 'seal' glazing reveals to an extent to minimize the amount of linseed oil in putty from soaking into the timber and, if used as a primer on top of putty, are required to prevent drying-out of the putty before subsequent coats of paint are applied.

Primers should contain a fungicide to prevent mould growth in and on the finish and, in the case of water thinned primers, to prevent putrefaction in the original container.

Note that certain of the exterior-quality alkyd and acrylic paints do not have separate primers but are applied 'coat-on-coat' to the timber. A separate conventional primer should not be used where such an exterior-quality finish is to be applied.

Note that a conventional primer should not be used if the top coat is to be a decorative stain.

It is important to ensure that any free solvent in organic solvent preservatives should be driven-off before the timber is primed. This is particularly important when complete factory finishing is involved. It is essential to ensure that the preservative is compatible with the paint (and any weatherseals).

Undercoat on exterior surfaces

A conventional or general-purpose undercoat *should not be used on an exterior surface*. The undercoat recommended for the top coat should be used. With some exterior-quality paint systems it is the same as the top coat.

Top coat on exterior surfaces

A general-purpose paint *should not be used on an exterior surface*.

A clear finish should only be used if regular maintenance (perhaps every year) is acceptable. With a clear finish on an exterior surface, photo-degradation will occur (with hardwoods as well as softwoods) at the interface between the timber and the finish.

There are five categories of top coats. They are:

Opaque exterior-quality alkyd paints

Often the top coat and undercoat (or undercoats) are identical and must be applied 'coat-on-coat'. Usually the finish is slightly microporous (if not applied too thickly) and will allow the passage of a certain amount of moisture vapour whilst acting as a barrier to rain. It must, however, be applied to a dry surface. Exterior-quality alkyds are suitable for application to smooth surfaces and normally have a satin or low-gloss finish rather than a high-gloss finish.

Opaque special exterior-quality formulations (e.g. alkyds or two-pack urethanes) intended only for factory finishing

The instructions of the paint manufacturer must be followed closely. Because the timber can be painted under ideal conditions with end grain being sealed, there is less need for the paint film to be microporous.

Opaque exterior-quality acrylic-emulsions

Although used on smooth surfaces (either for original painting or re-painting an exterior-quality alkyd), and on the top of horizontal members, they usually perform better on the type of fine sawn surface one associates with timber cladding, and perform better on vertical surfaces. (Indeed, this can be said to apply to more than acrylic emulsions.) Because they are water based and are fairly permeable, they will tolerate being applied to timber where the moisture content is slightly on the high side (although obviously application to a dry surface is preferred). They must be applied to an adequate thickness.

They are slightly thermoplastic, therefore they may be 'tacky' in warm weather and are likely to become dirty quicker than an alkyd gloss paint. 'Blocking' of surfaces which come into contact during manufacture can be a problem. They are available in a range of gloss levels depending on the formulation.

The paint should contain a fungicide to prevent mould growth in and on the paint and putrefaction in the original container. Even if the paint contains a rust inhibitor it is better to use rustproof or rust-proofed fittings.

If putty glazing is used, the paint manufacturer must be consulted to state if the paint is adequate to prevent drying-out of the putty. A test is included in BS5082: 1986 to require primers to be suitable when used to prime reveals.

Non-opaque exterior wood stains

There are many variations of wood stains with different degrees of opacity depending on the degree to which they are 'low-build' or 'high-build'. Although they are relatively easy to maintain if maintained in time, their life is usually shorter than that of an exterior-quality opaque finish, and therefore it is probably unwise to use them where access is difficult or expensive. If too long a period is left before maintenance is carried out, surface preparation for maintenance can be unnecessarily expensive. Generally lighter colours have a shorter life than dark colours and give less protection to the timber.

Stains having a film thickness of up to about 40 microns are more appropriate for use on an item such as vertical timber cladding than on exterior joinery such as a timber window. Stains with a film thickness of 20 microns or less have very little effect on moisture control, hence very little effect on the stability of the surface. Although low-build stains can be used on a planed surface, they perform better on a fine sawn surface. They have a much longer life on a vertical surface than on the top of a horizontal member, therefore protrusions (such as sills) should be kept to a minimum (and sloped). With some formulations, the life of the finish on top of a horizontal member can be extended by the use of an extra coat (and, if possible, by increasing the slope of the surface to shed water quickly).

Stains with a film thickness in excess of 40 microns are usually intended to be used on a smooth surface, but have shorter life than a thicker exterior-quality opaque paint.

Stains are unlikely to be suitable for use with putty glazing.

Clear finishes such as a varnish

A clear finish will not completely filter out ultraviolet light and therefore, if used in an exposed position, will require frequent maintenance. Some colour change will occur and photo-degradation will commence at the interface between the timber and the finish. It is essential that any break in the film which moisture can reach is repaired quickly, or failure will be accelerated and the timber will degrade local to the break. No clear finish is likely to be suitable with putty glazing.

ADDITIONAL NOTES ON PAINTING

Resin exudation

If the specification of the timber permits, resin pockets may be cut out and made good if this is compatible with the finish. If not, resin pockets may have to be excluded from the specification, or resin may have to be left to bleed through the finish and be wiped-off or knifed-off when dry. Generally, with a permeable/microporous finish, small amounts of resin such as can be encountered from around a knot will emerge without lifting off the finish.

At positions (e.g. around knots) where it looks as though resin may emerge at a later stage, the use of 'knotting' or a special primer may retard resin exudation. If shellac knotting is used under exterior surfaces which are largely permeable, a quality product which is not water sensitive should be chosen. Shellac is probably unsuitable for use under any non-opaque finish because it affects the degree of absorbency of the surface, hence will affect the local appearance of the finish. The paint manufacturer should be consulted about any special knotting or primer which may be available to retard resin exudation.

The solvent used in preserving timber with an organic solvent process may accelerate initial resin exudation.

Due to the non-reflective nature of dark colours leading to a hot surface in warm weather, the tendency for resin exudation is increased or accelerated if a dark finish is applied on an exterior timber surface, particularly if on an elevation which receives a great deal of sunlight.

Filler

Filler should be of high quality and have suitable durability.

Where filler is used under a non-opaque finish, it is necessary to choose one for its colour match to the timber, and with a surface absorbency which will not highlight the filler under the finish. Even with care, the filler is likely to show through a low-build stain. Particularly with a sawn surface, in many cases it may be preferable for appearance not to use a filler under a wood stain, but to accept that 'natural' characteristics are one of the features of a stained finish. Experience in practice shows that checks of 0.5 mm width or less cannot be filled.

Re-painting exterior surfaces

When re-painting, it is quite likely that exterior surfaces will have a slightly higher moisture content than is ideal. Because of this, it is almost certain that a finish with a degree of permeability will be required when re-painting. If the timber has been revealed for some time by damage to the original finish, the timber will have degraded ('greyed'), and it will be necessary to sand or plane the surface until clean timber is revealed before re-painted is commenced. On the other hand, UK experience points to it not being cost-effective to remove existing paint which is adhering well, but simply to prepare the surface for re-painting (e.g. by sanding).

Glazing details must be checked and any repair carried out. Paint will not fill gaps between putty and glass with any degree of permanence.

Reducing hazards to paint on exterior surfaces

The main hazards to paint on an exterior surface are rain and other moisture (e.g. from condensation), sunlight, and dust etc. in the wind and atmosphere.

The life of a paint finish can be increased by good detailing, choice of a suitable colour etc. Hazards to the paint film will be reduced by observing the points listed below (many of which are relevant to timber windows), some of which are illustrated in sketches in Fig. B7.1.

- Where possible, detail the building or component to shed moisture quickly and to reduce the amount of water and moisture which can reach the paint.
- Control the moisture content of the timber before painting: 15% moisture content or lower is to be preferred.
- The top of any horizontal exterior surface (and some horizontal interior surfaces) must be sloped and the slope should be as steep as possible. The more permeable the paint the more important it is to have a slope sufficiently steep to shed water quickly.
- The protrusion of the top of horizontal exterior surfaces should be

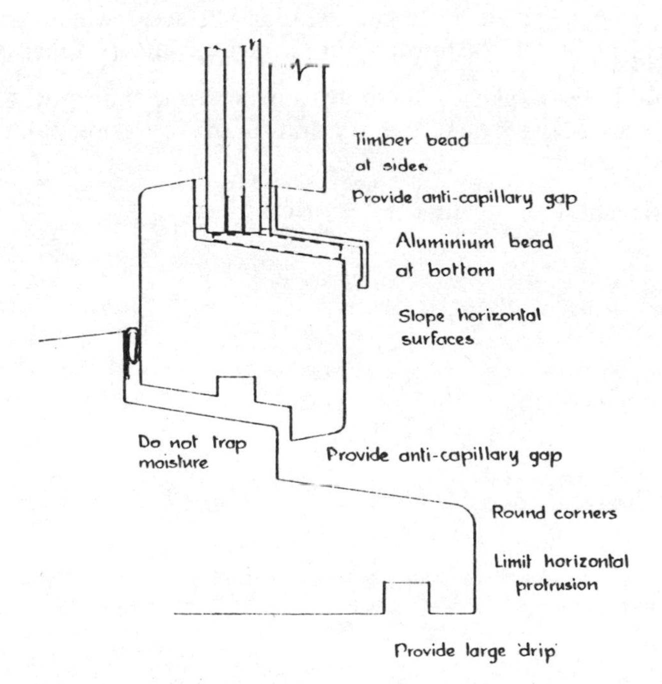

Fig. B7.1 Details of timber windows to minimize hazards to the paint on exterior surfaces

kept to a minimum. This is particularly important if a permeable finish is used.

- All surfaces which can be affected by rain must be detailed to shed water quickly and must be 'aired' so that, after rain has stopped, they dry quickly.
- Moisture traps must be eliminated. Where relevant, anti-capillary gaps should be provided.
- On windows, weatherseals must be fixed well back from the front surface. Preferably painting should be completed before weatherseals are fixed; if this is not possible, any paint should be wiped-off the seal as quickly as possible.
- The finish must be compatible with the preservative and fittings. Use rustproof or rust-proofed fittings.
- Damp-proof courses must be provided to prevent moisture ingress from the building (e.g. from walls) into the timber.
- Provide large 'drips' under horizontal members.
- If possible, seal end grain and any areas of high-porosity or, at least, minimize exposed end grain.
- Machine a 'round' (at least of 3 mm radius and preferably larger) on every corner (particularly on exterior surfaces) which is to be finished, and apply the finish to the correct thickness on each corner.
- Preferably use a glazing method more durable than putty but, if putty is used, use a quality putty and run the covering paint 2 mm onto the glass.
- Provide ventilation behind exterior cladding.

MECHANICAL FASTENERS

B8

INTRODUCTION

The most common types of mechanical fasteners for softwood are described in this chapter with brief notes on their use. When used in structural components, the requirements for end and edge distances and centres of fasteners, and the permissible loads per fastener are detailed in structural design code BS5268: Part 2: 1988.

NAILS

Wire nails are normally manufactured from mild steel. The relevant British Standard is BS1202: Part 1: 1974.

Round wire nails are used for carrying fairly small loads in shear (i.e. lateral load) along or across the grain, either in a timber-to-timber joint or a joint between timber and plywood, tempered hardboard, steel etc. Nails can carry (reduced) shear if inserted into end grain. The permissible withdrawal loads (in tension) are small if nails are driven into side grain, and zero if driven into end grain.

Nails can also be used for holding glued/nailed joints together whilst the adhesive is curing, or for permanent location of components even where the load is actually taken by timber-to-timber bearing, as in a timber stud wall.

With the exception of Hemlock (or Hem-fir), the softwoods covered by this book nail easily. Hemlock is prone to split if nailed when dry, although this splitting can be minimized by using nails with the point removed, or by pre-drilling.

For reasonable cost and delivery periods, it is important to specify or allow the manufacturer to use stock sizes. Stock sizes are usually up to 6 mm diameter and up to 150 mm in length, but it is necessary to check with a stockist. In certain conditions, rustproof or rust-proofed nails should be used. Their availability may be more limited than 'bright' nails (i.e. nails with no finish).

Oval wire nails or lost head wire nails are useful for joinery or

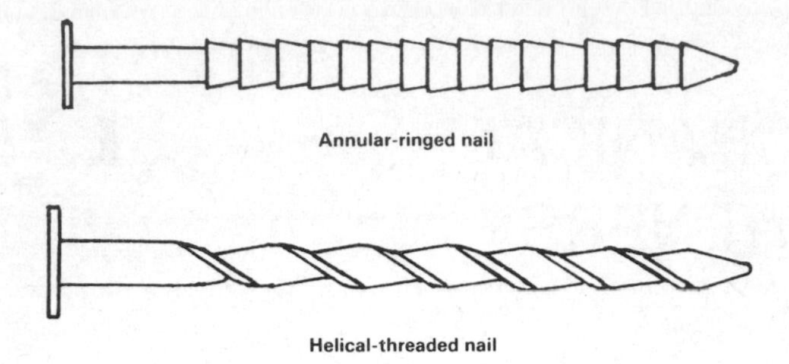

Fig. B8.1 Improved nails

carpentry where structural design of the joint is not required. 'Cut nails' or 'flooring brads' are used for fixing softwood floor boards.

IMPROVED NAILS

Improved nails such as annular-ringed (ring-shanked) or helical-threaded nails (see Fig. B8.1), or occasionally nails with an unthreaded square or cruciform cross-section are available unprotected or rust-proofed. Most are manufactured from mild steel. The range of sizes is more limited than that for ordinary wire nails.

A ringed or threaded nail is useful to give added resistance to withdrawal during construction or in service, or to carry a larger tension design load than an ordinary wire nail. A nail with a square cross-section, either straight or twisted, is useful to carry a larger lateral load than a round nail.

STAPLES

Staples can be used to take lateral load between a sheet material and solid timber, either for permanent use (e.g. between the plywood and studs in a timber stud wall), or for a temporary use although left in place (e.g. to give close contact during curing of a glue line). Staples are nearly always inserted by a stapling 'gun' which must be adjusted not to leave the staple clear of the surface nor to fire it too far into the surface, particularly if fired into plywood. (The same warning is relevant if nails are inserted by a nailing gun.)

Staples are available in mild steel or other metals in several combinations of length, width and gauge. There are differing opinions as to whether or not it is preferable to use 'divergent point' staples, and a certain amount of discussion with the manufacturer, and/or experimenting for a specific use is recommended.

SCREWS

Screws can be extremely useful in pulling two pieces together, by providing a joint which is easy to dismantle, or in providing more resistance to withdrawal than a nail or improved nail. It should be realized, however, that a nail of the same nominal diameter as a screw has more resistance to lateral load, and that a screw is expensive to insert, particularly if pre-drilling and countersinking is necessary. The relevant British Standard is BS1210: 1963.

When screws (or coach screws) are used structurally, structural design code BS5268: Part 2: 1988 requires them to be inserted into pre-drilled holes of a specified diameter.

The most common type of screw is the countersunk head steel screw of which many stock sizes are available, normally up to 8.4 mm diameter and up to 100 mm length. Rustproof or rust-proofed screws are available, as are round headed screws.

COACH SCREWS

For special uses requiring large screws, mild steel 'coach screws' are available. The largest diameter normally readily available is 12 mm, in lengths up to 150 mm. Larger and longer coach screws usually need to be ordered specially. Coach screws are available rust-proofed. They are turned by a spanner rather than a screwdriver. They need to be inserted into pre-drilled holes.

BOLTS

Bolts as used in timber engineering connections tend to have a large length/diameter ratio (e.g. lengths up to 400–500 mm with a diameter of 20 mm). It is prudent to check the availability before specifying. If necessary, bolts can be rust-proofed, in which case the specifier or manufacturer must ensure that the rust-proofing is carried out in such a way that the nut can be turned on the thread of the bolt without the rustproofing being stripped off. It may be necesarry to 'tap' out the nut or 'run down' the thread on the bolt before rustproofing takes place. The relevant British Standard is BS4190: 1967.

Bolts in timber engineering normally require the use of relatively large and thick washers under the head and nut. The relevant British Standard is BS4320: 1968. The washers used in timber engineering are normally three times the diameter of the bolt with which they are used. If appearance is important, consideration should be given to the use of round rather than square washers because square washers, if used, may have to be lined up for appearance, which is difficult and time-consuming to achieve.

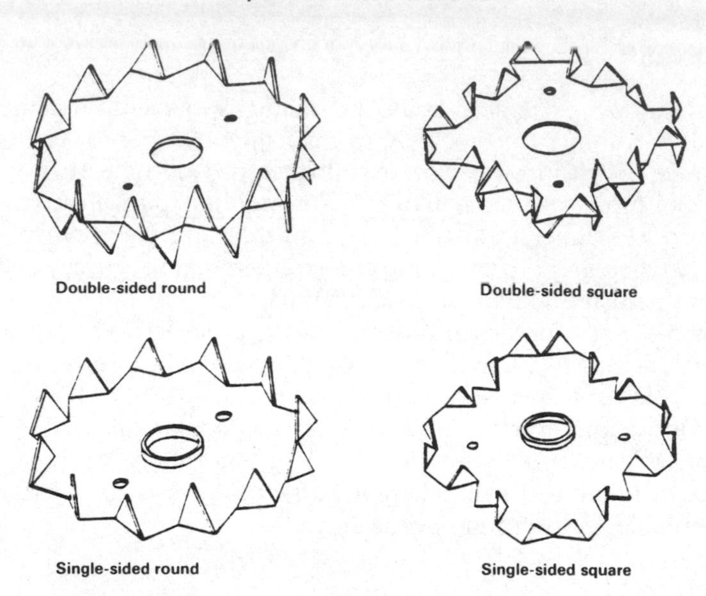

Fig. B8.2 Toothed-plate connector units

If a bolt is driven into oversize holes and the connection is in shear, slip will take place which is likely to lead to slip deflection of the component (e.g. of a truss).

TOOTHED-PLATE CONNECTOR UNITS

Toothed-plate connectors are round or square, in sizes up to 76 mm, and are used in conjunction with a bolt of a specific diameter. They are normally supplied rustproofed. As shown diagrammatically in Fig. B8.2, they are available single sided or double sided and can be used in single shear or double shear. The larger sizes may well need to be drawn into the timber by a high tensile 'drawing' stud. The relevant British Standard is BS1579: 1960.

Each plate has a thickness of about 2 mm which means that the surfaces of the timber components being joined are not in close contact. If single-sided connectors are used, slip may take place.

SPLIT-RING CONNECTORS

Steel split-ring connectors are available in two sizes which are 64 mm and 104 mm nominal diameter. A diagrammatic cross-section is shown in Fig. B8.3 of one type which shows the 'split' by which the ring is able to move slightly in assembly and service. Split-rings are used in conjunction with a bolt and washers to hold the timbers together (see

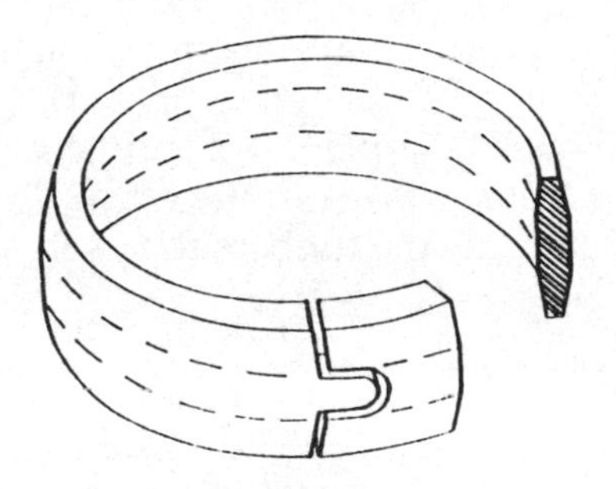

Fig. B8.3 Split-ring connector

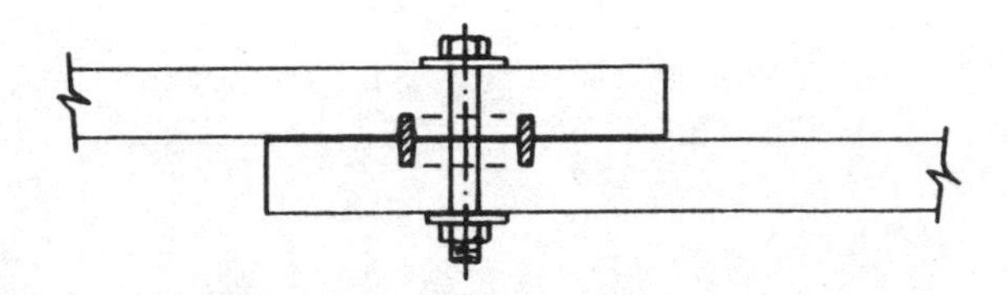

Fig. B8.4 Assembly of split-ring connector unit

Fig. B8.4). A special tool is required to groove out the timber for the shape of the split-ring, and a special 'drawing' tool is required to assemble the unit under pressure.

Split-ring connector units are useful where a relatively high lateral load is to be carried.

The relevant British Standard is BS1579: 1960.

SHEAR-PLATE CONNECTORS

Shear-plate connectors are available in two sizes which are 67 mm and 102 mm outside diameter as sketched in Fig. B8.5. The smaller is made

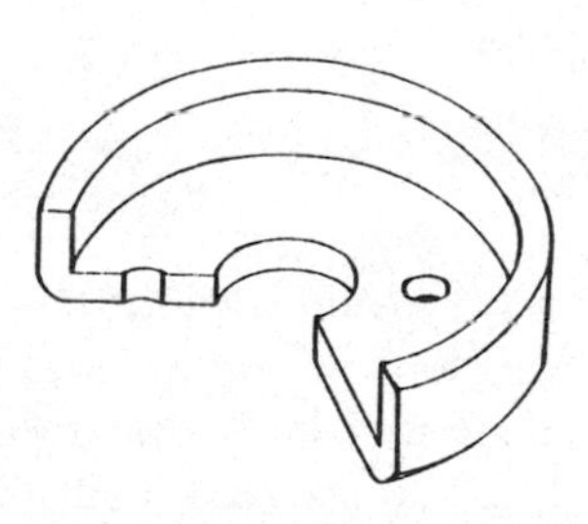

67 mm pressed steel shear-plate connector

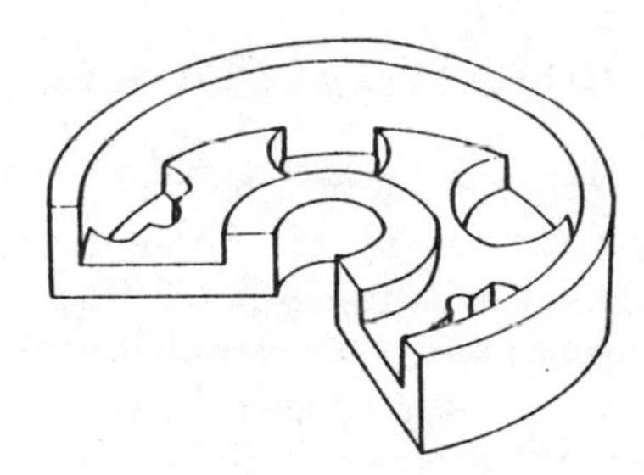

102 mm cast iron shear-plate connector

Fig. B8.5 Shear-plate connectors

from pressed steel, the larger from cast iron. Diagrammatic examples of uses are shown in Fig. B8.6. A shear-plate is used with a bolt and washers. Slight slip can take place before the bolt is in bearing. A special tool is required to groove out the timber for the connector but no tool is required to draw the timbers together.

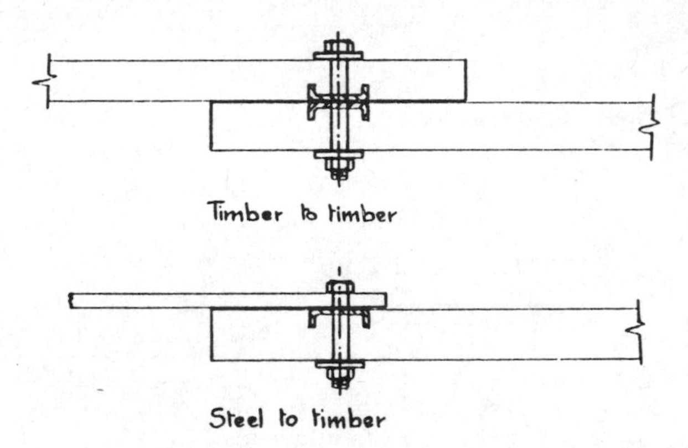

Fig. B8.6 Typical assemblies of shear-plate connector units

Shear-plate connectors are easier to use than split-ring connector units if the components are to be manufactured in a factory for site assembly. As can be seen from Fig. B8.6, the surfaces of the timber components which are joined are in close contact. Shear-plate connectors are held in place for transit by two small locating nails. The pressed steel shear-plate connector must be given an anti-corrosion treatment.

The relevant British Standard is BS1579: 1960.

PUNCHED METAL PLATE FASTENERS

Thin proprietary punched metal plate fasteners with integral teeth have proved to be very popular and successful as gusset plates in the manufacture of triangulated frameworks such as lightweight trussed rafters and lattice beams. Slight slip occurs as load is applied and this must be considered in any design, particularly in a lattice beam or a trussed rafter with a shallow slope. Plates must be inserted (pressed in or rolled in) by purpose-built equipment. Plates are inserted in pairs, one on each side of the timber component.

Most plates are of zinc coated 18–20 gauge steel plates, but stainless steel plates are available at extra cost. Zinc coated plates normally have the teeth punched out and are cut to size after being coated, therefore there are many edges where the steel is exposed. Despite this, in normal building or roof environments with normal relative humidity, adequate ventilation and properly dried timber, corrosion has not been found to be a problem even when the components have been subjected to a certain amount of rain during delivery, site storage and erection. If corrosion commences on an exposed edge of a thin zinc coated plate, the zinc limits corrosion by performing in a 'self-sacrificial' manner across the narrow gap between the treated faces.

Adequate ventilation must be provided to ensure that the moisture content of the timber does not increase in service. Some concern has been expressed about the possibility of accelerated corrosion where plates are inserted into timber treated with a CCA water borne preservative. Manufacturers of metal plated trussed rafters, if called upon to preserve, prefer to use an organic solvent treatment which does not add moisture to the timber. If, however, there is no option but to use a water borne treatment and zinc coated plates, the timber must be re-dried to around 18% moisture content or lower before the plates are inserted. As stated above, adequate ventilation in service must be provided.

Metal plates with pre-punched holes for the insertion of nails are available, and offer a useful alternative to metal fasteners with integral teeth where purpose built assembly equipment is not available.

STEEL GUSSETS AND HANGERS

Steel gussets, shoe plates etc. can be used in timber connections in a similar way to how they are used in structural steelwork connections. Depending on the magnitude of the load to be transmitted to or from the timber, nails, screws or bolts can be used with steel gussets or hangers.

Proprietary joist hangers etc. or purpose-made steel hangers, brackets etc. can also be used. If a tight fit is required between timber and steel (e.g. for a compression joint or to prevent moisture ingress), an epoxy resin/sand mix or similar can be used to cater for tolerances in the timber or steelwork.

The finish to the steelwork and fasteners must be suitable for the environment.

OTHER FASTENERS

Ragbolts, Rawlbolts, ballistic nails etc. all have their uses with structural timber, as have many special fasteners such as dowels designed for joinery.

COMPATIBILITY

The material of the fastener and/or its finish must be compatible with any preservative or finish to the timber, and with the service environment. CCA preservatives have been found to react with uncoated aluminium fasteners. In special cases, it may be necessary to use stainless steel or phosphor-bronze fasteners. Galvanized fasteners should not be used in timber which has been treated with a flame retardant, unless the formulation is a non-salt type.

Western Red Cedar and Douglas Fir-Larch are quoted in BS1186: Part 1: 1986 as being acidic timbers and, if unprotected metal is in contact with them, it is stated that accelerated corrosion can take place leading to a black 'iron stain'. The UK Building Research Establishment advise the use of hot-dip galvanizing or copper nails for fasteners in these timbers.

GLUED JOINTS B9

TYPES OF ADHESIVE

The types of adhesive most commonly used for structural use and joinery use when gluing softwood to softwood or wood based panel products are:

(1) For structural uses:

resorcinol-formaldehyde (RF),
phenol-formaldehyde (PF),
phenol/resorcinol-formaldehyde (PF/RF),
melamine/urea-formaldehyde (MF/UF),
other modified urea-formaldehyde adhesives,
urea-formaldehyde (UF),
casein.

The first six adhesive types listed above are thermo-setting resin adhesives which, for structural use, should comply with BS1204: Part 1: 1979 *Synthetic resin adhesives for wood. Specification for gap-filling adhesives.*

Casein for structural use should comply with BS1444: 1970 *Cold-setting casein adhesive powders for wood.* (This standard has been withdrawn despite objections from industry. It is referred to in a few current British Standards.)

(2) For joinery uses, any of the adhesives listed above plus:

catalysed polyvinyl acetate,
polyvinyl acetate (non-catalysed),
special formulations for special purposes.

Currently there is no British Standard for catalysed polyvinyl acetate adhesives although BS4071: 1966 covers *Specification for polyvinyl acetate emulsion adhesives for wood.* Although a Draft for Development DD74: 1981 *Performance requirements and test methods for non-structural wood adhesives* was published by BSI, it is more common in the UK to specify polyvinyl acetate adhesives to one of the levels of German DIN Standard 68 602: 1979 *Evaluation of adhesives for jointing of wood and derived timber products. Strain groups. Strength of bond.* Polyvinyl acetate adhesives are thermo-plastic and should not be used for structural purposes. They are not formulated as gap-filling adhesives. (BRE Digest 340: 1989 provides useful reading. It supercedes BRE Digests 175 and 209 which have been withdrawn. It points out that catalysed polyvinyl acetate adhesives are normally two-part formulations whilst non-catalysed are one-part, but comments on the more recent development of one-part formulations which exhibit a certain degree of cross-linking when they cure.)

Other adhesives may be used (e.g. epoxy resin) for particular applications. Special modifications may be required for certain applications. For example, some adhesives based on resorcinol resin have a stone-flour additive when used for finger jointing to reduce the amount of adhesive pressed into the end grain at time of assembly.

DURABILITY

BS1204: Part 1: 1979 and BS1204: Part 2: 1979 give the following durability ratings and the tests required to match these ratings:

Type WBP: Weather-proof and boil-proof

Adhesives of the type which, by systematic tests and by their records in service over many years, have been proved to make joints highly resistant to weather, micro-organisms, cold and boiling water, steam and dry heat.

Type BR: Boil-resistant

Joints made with these adhesives have good resistance to weather and to the test for resistance to boiling water, but fail under the very prolonged exposure to weather that Type WBP adhesives will withstand. Joints made with these adhesives will withstand cold water for many years and are highly resistant to attack by micro-organisms.

Type MR: Moisture-resistant and moderately weather-resistant

Joints made with these adhesives will survive full exposure to weather for only a few years. They withstand cold water for a long period and hot water for a limited time, but fail under the test for resistance to boiling water. They are resistant to attack by micro-organisms.

Type INT: Interior

Joints made with these adhesives are resistant to cold water but are not required to withstand attack by micro-organisms.

(Note: One does encounter some adhesives being cross-referenced to BS1204: Part 1 even when they are not resin adhesives.)

The durability ratings in Part 2 are the same as in Part 1, but close-contact adhesives are not normally suitable for, nor used for structural purposes.

Still on the subject of durability in service, BS5442: Part 3: 1979 gives a classification of the suitability of adhesives (not just resin adhesives) for several uses in building and construction. The codes are:

A Suitable for service inside heated and ventilated buildings and in other environments where the equilibrium moisture content of natural softwood does not exceed 18% and the bond line remains below a temperature of 50°C.

BC Suitable for exposure conditions intermediate between A and D bearing in mind that any given set of exposure conditions can have different results depending upon the type of wood and on the article or fabrication in question.

D Suitable for full exposure to weather or to water immersion.

S Could be suitable in several cases, but is not used in general practice.

The adhesives listed at the beginning of this chapter are considered to fit into the durability ratings of BS1204: Part 1: 1979, and/or the classifications of BS5442: Part 3: 1979 as shown in Table B9.1.

HAZARD CONDITIONS FOR STRUCTURAL ADHESIVES IN BUILDINGS

In addition to the guidance given in BS5442: Part 3: 1979, BRE Digest 175: March 1975 (now withdrawn) gave a table of guidance on the choice of adhesives for different applications which, in turn, is used as a basis for similar tables in structural design code BS5268: Part 2: 1988 and in BS6446: 1984 on the manufacture of glued structural components of timber and wood based panel products. An edited version is presented in Table B9.2.

Table B9.1 Durability rating of BS1204: Part 1: 1979 and classification of BS5442: Part 3: 1979

Glue type	Durability rating of BS1204: Part 1: 1979[1]	Classification of BS5442: Part 3: 1979
RF	WBP	D
PF	WBP	D
PF/RF	WBP	D
MF/UF	BR[2]	BC[3]
Other modified UF	BR[2]	BC[3]
UF	MR[3]	BC[3]
Casein to BS1444	INT (not a resin adhesive)	A
Catalysed polyvinyl acetate	[4]	BC
Non-catalysed polyvinyl acetate	INT	A

[1] See text for notes on BS1204: Part 2: 1979.

[2] It is necessary to check each formulation for its suitability for the service conditions, and ability to match BR.

[3] Certain urea-formaldehyde-based adhesives which comply with the requirements of Type MR of BS1204: Part 1: 1979 and Part 2: 1979 have been found suitable for non-structural applications where the moisture content of softwood can exceed 18% in service but which does not remain at this figure continuously, where joints are protected by paint films and are not subjected to loads. These adhesives are generally found to give satisfactory service in these situations where the temperature may exceed 50°C but each formulation needs to be checked. (Where UF-based adhesives are used in windows or external doors, BS5442: Part 3: 1979 draws attention to the need to choose a formulation which can match the requirements of moisture content and temperature.)

[4] Some catalysed polyvinyl acetate formulations have been found to pass the WBP test level of BS1204: Part 1: 1979. However, there are doubts about the long-term performance of such adhesives where the moisture content can be as high as 18% or more for periods of time, and/or where the glue line temperature approaches that which can be taken by a WBP adhesive (i.e. 50°C). It is said that there is no correlation between the results of the tests for WBP in BS1204: Part 1: 1979 and the long-term performance of catalysed polyvinyl acetate adhesives in situations exposed to weather or moisture. At this stage in the experience of catalysed polyvinyl acetate it is advisable to consult the adhesive manufacturer when employing a formulation for uses such as the laminating of window sills, particularly if the timber is to be covered by a dark coloured paint or a decorative stain. For example, by designing a sill so that the glue lines are kept well back from the outermost face, hazard to the glue lines is reduced.

QUALITY CONTROL

To produce reliable glue lines consistently, routine quality control is necessary. Various codes and standards give the actual levels of permitted moisture content etc. for various components. The method of manufacture and curing also influences the required quality control. In general the aspects which must be controlled are:

- The moisture content of the timber (and wood based panel products) at the time of gluing and curing.
- The temperature of the timber and the surrounding air at the time of gluing, and during storage before and after gluing.

Table B9.2 Structural adhesive types for exposure categories in buildings

Exposure category	Exposure conditions	Examples	Recommended adhesive	Performance type and BS reference
Exterior				
High hazard	Full exposure to weather	Marine and other exterior structures. Exterior components or assemblies where the glueline is exposed to the weather[1]	RF PF/RF PF (cold-setting)	WBP BS1204: Part 1
Low hazard	Protected from sun and rain	Inside the roofs of open sheds and porches	RF RF/PF PF	WBP BS1204: Part 1
			MF/UF[2] Some other modified UF adhesives[2]	BR BS1204: Part 1
Interior				
High hazard	In buildings with warm and damp conditions where a m.c. of 18% is exceeded and/or where the glueline temperature can exceed 50°C	Laundries, unventilated roof spaces. Chemically polluted atmospheres e.g. chemical works, dye works and swimming pools. External single leaf walls with protective cladding	RF PF/RF PF	WBP BS1204: Part 1
Low hazard	In heated and ventilated buildings where the m.c. of the wood will not exceed 18% and where the temperature of the glueline will remain below 50°C	Interiors of houses, buildings, halls and churches, inner leaf of cavity walls	RF PF/RF PF	WBP BS1204: Part 1
			MF/UF[2] Some other modified UF adhesive[2]	BR BS1204: Part 1
			UF[2]	MR BS1204: Part 1
			Casein	BS1444

[1] Glued components other than glued laminated members are not recommended for use at this exposure category.

[2] Care must be taken to ensure that a particular formulation is suitable for the service condition and the intended life of the structure.

- The application and curing of the adhesive.
- Protection from weather, cold draughts etc. during full-cure (even if accelerated curing is used).
- Storage and mixing of adhesives, glue tests etc. (all to the instructions of the adhesive manufacturer).
- The quality of surfaces to be glued and the need to restrict the time between machining and gluing.
- Handling of sub-components and components after gluing.

For certain structural components it may be necessary to provide a further check on the quality of the glue line by carrying out routine tests to destruction on small samples typical of the production. Some British Standards such as BS6446: 1984 require routine tests to destruction as part of the quality control.

CURING OF GLUE LINES

Throughout the curing period, the glue line must be kept under uniform pressure by clamps, nails etc. If accelerated curing is used, temporary clamps etc. may be removed earlier than otherwise, but the component will still require careful handling during the remaining period of 'full-cure' (perhaps seven days or longer). If accelerated curing is not used, any temporary clamps etc. must be left undisturbed during the full curing period unless designed only to hold the assembly together until permanent fixings such as nails are inserted. If radio-frequency accelerated curing is used, nails etc. must not be used (other than perhaps for location purposes if this will not lead to trouble). The viscosity of newly mixed adhesive reduces considerably at high temperature, and a metal fixing acts as a short-circuit path.

Some typical examples of glued components are sketched in Table B9.3 with brief notes of the typical type of fastener or process one can expect to be used to ensure close contact or pressure during curing of the adhesive.

COMPATIBILITY

If the timber is preserved, the adhesive must be compatible with the preservative, and the manufacturers of the adhesive and preservative should be consulted to see if gluing can take place after preserving, or if preserving can take place after gluing. Some adhesives cannot be applied after timber has been preserved with a solution containing a water repellent, although satisfactory results have been obtained with

Table B9.3 Examples of the fixings or processes used to ensure close contact of glue lines during adhesive curing

Typical examples of glued components	Methods of applying pressure or ensuring close contact during curing
Structural glulam	Clamps, pneumatic pressure, purpose-designed automatic equipment, sometimes with accelerated curing
Ply-web beams	Normally nail pressure, although pneumatic pressure or clamps have also been used. (Normally excess glue will be visible in the final assembly.) The nails are left in place although inserted primarily to hold the glue lines during curing. One proprietory system uses pressure with no mechanical fasteners
Stressed-skin panels	Normally nail (or staple) pressure, the nails or staples being left in place although inserted primarily to hold the glue lines during curing
Scarf joints	Nail pressure or clamps

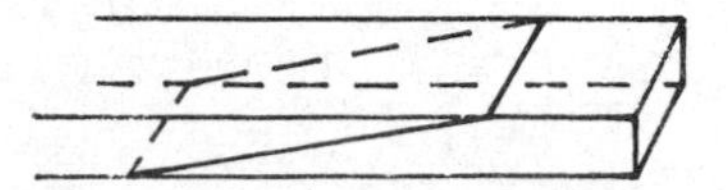

Glued plywood gussets in the assembly of trusses	Normally nail (or staple) pressure, the nails or staples being left in place although inserted primarily to hold the glue lines during curing. As an alternative, multiple assembly clamps have been used, pressure being maintained during curing

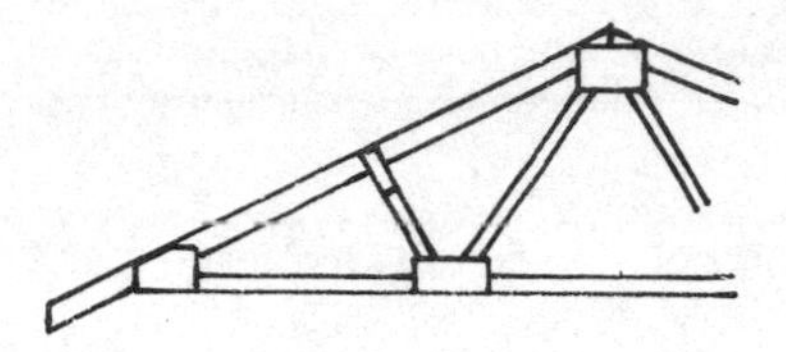

Finger joints	End pressure with no mechanical fastener (see Chapter B10). Accelerated curing or pre-heating of the timber may be employed

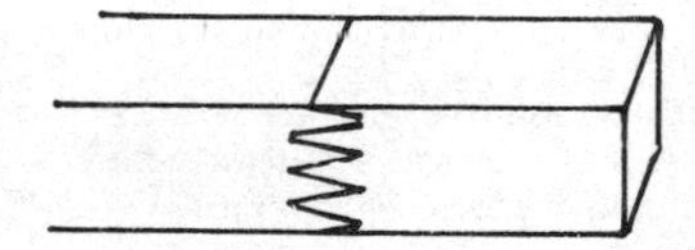

Joinery laminating	Clamps in purpose made equipment (no mechanical fasteners). Often accelerated curing is used

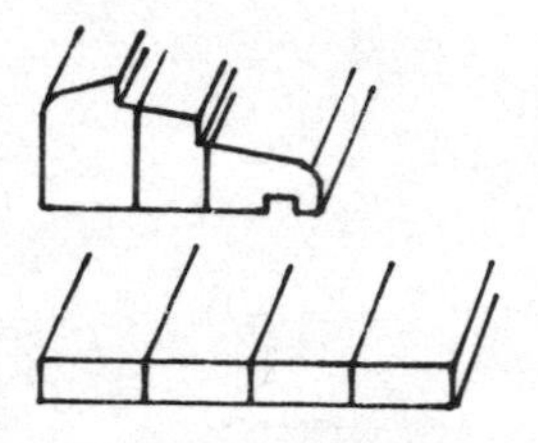

Framing for flush doors	Multiple-assembly presses (no mechanical fasteners). Often accelerated curing is used

Window corner joints	Nail or specialist fixing (e.g. star dowel)

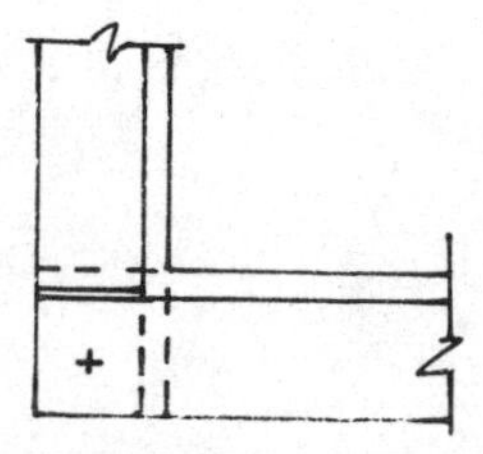

Table B9.4 DIN 68 602: 1979 category groups

Group	Requirements of the adhesive	Examples of typical uses
B1	Resistant in indoor areas with generally low humidity provided that the outdoor atmosphere cannot also have a direct influence on the temperature and humidity	Dry indoor areas (e.g. doors, furniture)
B2	Resistant in indoor areas with high and fluctuating humidity of brief duration and occasional brief exposure to moisture	In indoor areas with higher humidity (e.g. kitchens, bathrooms)
B3	Resistant in climatic influences in a 'moderate climate' (defined elsewhere in another DIN)	In indoor areas with increased humidity for short duration and brief exposure to water; and in some outdoor uses (e.g. doors, windows and exterior stairs)
B4	Resistant to climatic influences in a zone with 'moderate climate' (defined elsewhere in another DIN) under particularly unfavourable conditions	In indoor areas with extreme variations of climate and influence of water (e.g. swimming baths, shower cubicles), and in outdoor uses with highly adverse climatic influences (e.g. windows and outside doors covered with a decorative stain or dark coatings; ladders and exterior stairs).

certain adhesives and certain preservatives containing a water repellent.

If the timber has been treated with a flame retardant containing ammonia, it is not normally possible to glue afterwards. It may be possible to apply a flame retardant to a completed glued component if a WBP adhesive has been used, providing that full-cure is complete before treatment with the flame retardant is commenced.

GERMAN STANDARD DIN 68 602: 1979

Occasionally, for non-structural components, reference is made in the UK to Groups B1, B2, B3 or B4 of German DIN 68 602: 1979, therefore some extracts are given in Table B9.4.

The categories are called 'Stress Groups' or 'Strain Groups' (or perhaps the translation should be 'Strength Groups') because, in general, to qualify for a group, test pieces made from adhesives of a B1, B2, B3 or B4 category must exhibit prescribed shear strengths (as shown in Table B9.5) after the tests as outlined in Table B9.5 have been carried out. (Note that there may be slight changes of meaning due to translation.)

Table B9.5 Tests for DIN 68 602: 1979 stress groups

Strength of bond in N/mm² for stress groups				Brief description of test procedure The various 'climates' are defined in other DINs
B1	B2	B3	B4	
⩾10	⩾10	⩾10	⩾10	7 days in normal climate
—	⩾5	—	—	7 days in normal climate 3 hours in cold water[1] 7 days in normal climate
—	—	⩾2	⩾2	7 days in normal climate 4 days in cold water[1]
—	—	⩾6	—	7 days in normal climate 4 days in cold water[1] 7 days in normal climate
—	—	—	⩾4	7 days in normal climate 6 hours in boiling water 2 hours in cold water[1]
—	—	—	⩾8	7 days in normal climate 6 hours in boiling water 2 hours in cold water[1] 7 days in normal climate

Where a gap occurs in the table, no test is required. To satisfy the B1 Stress Group, one test is required. To satisfy the B4 Stress Group, four tests are required. Each test is carried out after the total prescribed 'storage' sequence given in the right hand column.

[1] 20°C ± 2°C.

FINGER JOINTS B10

INTRODUCTION

Although some specifiers still regard the process of finger jointing as being relatively new, it has been used successfully since the mid-1940s for joinery and structural uses and is accepted in major British Standards dealing with joinery and structural uses. Finger joints are self-locating glued end joints in which a number of tapered symmetrical fingers are cut by purpose-made machinery in the ends of the timber members to be joined. The fingers are then glued and the joints are assembled under controlled axial pressure. Typical profiles are shown in Fig. B10.1. Although variations on profiles and different forming methods have been developed, in practice profiles cut as shown in Fig. B10.1 have been found to give the strengths and function required of a joint. The profiles can be cut to be visible on either the broad or narrow face of a section without strength being affected, provided that at least four fingers occur on the face on which the profile is visible.

As an alternative to cutting the finger profiles, short finger joints in the order of 3 mm long have been produced by pressing the square ends of timber against a die (i.e. a die-impressed joint). Slightly longer joints have been produced by a combination of die-impression and cutting. Only cut joints are dealt with in this chapter.

Unlike a gusset-type end joint, the joint is contained within the cross-section of the timber (which is an advantage during delivery) and, unlike a scarf joint, each joint is relatively short therefore does not waste long lengths of timber.

Normally a finger joint is used to join lengths of timber either:

- because a defect has been cut out to upgrade quality, or
- because longer lengths are required than are available from the lengths sawn from a log, or
- to recover short lengths.

In special cases, finger joints can be introduced at close centres to add to the dimensional stability of a section. Finger joints have been used as the basis for corner joints but that use is not dealt with in this book.

The principle British Standard on the manufacture of structural finger joints is BS5291: 1984. The clauses on the manufacture of finger joints for joinery are contained in BS1186: Part 2: 1987.

PROFILES, USE AND STRENGTH

In describing profiles, it is convenient to think in terms of three types of joint which are:

- the longer structural joints,
- the longer joinery joints, and
- shorter joints.

The early structural finger joints were made relatively long (as shown at A in Fig. B10.1) because developments had tended to indicate that providing long fingers, and keeping the angle of the fingers as close as possible to the longitudinal axis of the timber, was the way to achieve maximum strength. The 50 mm long joint was, and still is popular in the UK. The strength of a joint in bending and tension depends on the shearing strength of the various glue joints on the sloping surfaces of fingers. To ensure, therefore, that bearing does not take place on the tips of these relatively long joints during assembly before there is tight fit on the sloping surfaces, a gap is deliberately introduced at the tips. With this type of joint, bending and tension strengths of up to about 75% of the strength of defect-free timber can be achieved. In compression, the joint has hardly any strength reducing effect at all. Although suitable for many structural uses, it is obvious that the gap at each tip makes this type of joint unsuitable for most joinery uses, therefore a joinery joint was developed of the type shown at B in Fig. B10.1.

The joinery joint shown at B has a relatively large angle between the sloping surfaces of the fingers and the longitudinal axis of the timber,

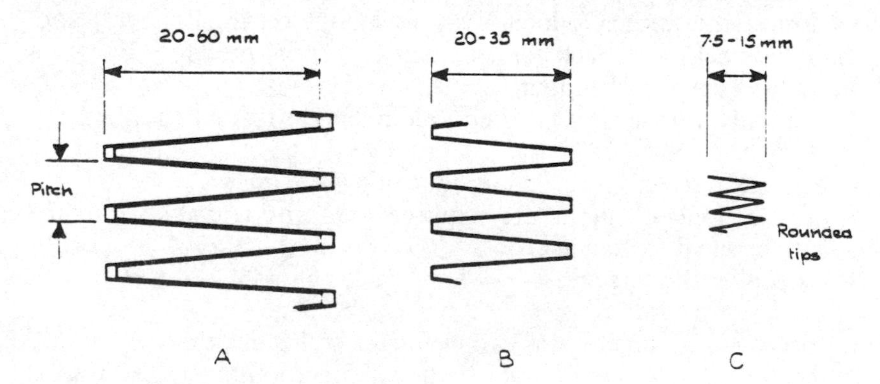

Fig. B10.1 Typical profiles of finger joints

and it also has a relatively wide tip width. This gives a reasonable certainty that, during manufacture, close contact can be achieved simultaneously on all surfaces. The absence of gaps at the tips makes this type of joint suitable for joinery, but it is not as strong as the structural joint shown at A. Strengths can be as low as 30–40% of the strength of defect-free timber.

Around the 1960s, it was established that the strength of a joint is more dependant on the angle of the fingers and the *total* length of glue line rather than the length of individual fingers, and this knowledge opened the way for the development of 'mini-joints' such as those shown at C in Fig. B10.1. Because the cutters are shorter, the accuracy of cutting is increased and, because the fingers are short, the longitudinal assembly pressure can be much higher than that associated with longer joints without causing the timber to split. Although joints as short as 3 mm have been cut (or formed by die-impression), the optimum length for strength and manufacturing criteria has been found to be around 10 mm and 15 mm. Short joints have advantages, one being that they are unobtrusive, but perhaps the main advantage is that one profile (and cutters etc.) is suitable for joinery as well as structural use. As well as having adequate strength for structural uses, there are no discernible gaps at the tips. A 15 mm long joint can have a strength in bending and tension of as much as 75% of the strength of defect-free timber. In compression, the joint has hardly any strength reducing effect at all.

Table B10.1 Guide to the efficiency and use of finger joints and the countries where they are used

Length (mm)	Pitch (mm)	Tip width (mm)	Efficiency %[1]		Countries where known to be cut					
			Bending and Tension	Comp	UK	Sweden	Finland	Norway	Russia	Canada
55	12.5	1.5	75	88	*			*		
50	12.0	2.0	75	83	*		*		*	
40	9.0	1.0	65	89						
32	6.2	0.5	75	92		*		*		
30	6.5	1.5	55	77			*			
30	11.0	2.7	50	75						
20	6.2	1.0	65	84	+	*	*	*	+	
15	3.8	0.5	75	87	*	*	*	*		
12.5	4.0	0.7	65	82	*					
12.5	3.0	0.5	65	83	*					
10	3.7	0.6	65	84				+	*	
10	3.8	0.6	65	84	*	*	*			
7.5	2.5	0.2	65	92						

[1] Provided that at least four complete fingers are present.
*, Profile used.
+, Similar profile used.

This book is limited to the softwoods from Sweden, Finland, Norway, Canada, Russia, Poland, Czechoslovakia, Portugal, Brazil and the UK. Of these countries, finger joints are known to be used widely in Sweden, Finland, Norway and the UK, and are used for special purposes (e.g. glulam) in Canada with further developments currently under way. Finger jointing is reported to be used in Russia.

Finger jointing is used frequently in the UK in European Redwood and European Whitewood, Canadian Hem-Fir or Hemlock, Douglas Fir-Larch and British-grown Sitka Spruce, but there is little evidence of it being used for jointing Cedar, Parana Pine or Maritime Pine although there is no reason to suppose that these species cannot be finger jointed if the correct quality control is exercised.

The British Standard on structural finger jointing, BS5291: 1984, is quite specific in prohibiting the use of structural finger joints between different species, hence parcels of mixed species should not be finger jointed for structural use.

Table B10.1 lists some profiles and efficiencies and, where known, the countries where they are used or have been used. The tabulated efficiency ratings in bending and tension can be regarded as a comparison of the strength of finger jointed timber to unjointed defect-free timber. The efficiencies in compression are based on the formula given below from BS5268: Part 2: 1988:

$$\frac{(p - t)}{p} \times 100\%$$

where p is the 'pitch' of the fingers, and t is the width of the tips (see Fig. B10.1). Note that the strength of a finger joint in compression is always more than its efficiency in bending or tension.

In many cases (whether or not the timber is stressed to its full limit in bending and/or tension) it is easy to determine if the strength of a chosen finger joint is adequate for the stress grade of the timber. In these cases, all that the designer need do is compare the efficiency rating in bending for the stress grade of the timber which is required if the timber is fully stressed in bending (see Table B10.2 and the next paragraph), with the efficiency rating in bending of the finger joint (see Table B10.1). If the efficiency of the finger joint in bending is at least as high as the efficiency of the stress grade in bending, no further design check on the joint is necessary.

Table B10.2 gives finger joint efficiencies in bending required to satisfy the primary stress grades of BS5268: Part 2: 1988 when the timber is fully stressed in bending. As stated above, by comparing the values given in Tables B10.1 and B10.2, a specifier should be able to determine if a particular finger joint is strong enough for a particular stress grade without any further design check. Note that, however,

Table B10.2 Required finger joint efficiency if a stress grade is fully stressed in bending

Stress grade	Required finger joint efficiency (%) in bending if stress grade is fully stressed in bending	
BS4978 stress grades		
M75	75	
SS or MSS	60	
M50	50	
GS or MGS	50	Note that a finger joint of at least 50% efficiency is required by BS5268: Part 2: 1988
Glulam laminates from BS4978		
LA	—	Note that an LA laminate cannot be finger jointed without a strength penalty reducing its efficiency
LB	70	
LC	55	
Canadian NLGA stress grades		
Joist and Plank		
Select	65	
No. 1, No. 2, No. 3	50	
Structural Light Framing		
Select	65	
No. 1, No. 2, No. 3	50	Note that a finger joint of at least 50% efficiency is required by BS5268: Part 2: 1988
Light Framing, Utility and Stud Grade		
Construction	50	
Standard	50	
Utility	50	
Stud Grade	50	

even if this simple check does not show the finger joint to be strong enough for the full bending strength of the stress grade, a design check based on the actual stresses involved in the particular design case being investigated may still show the finger joint to have adequate strength. Such a check is permitted by BS5268: Part 2: 1988 and often timber is not fully stressed.

There are only a limited number of machine and cutter manufacturers, and a pooling of expertise between countries, therefore there is considerable harmonization of joint profiles etc. This has helped in the drafting of a European Standard for structural finger joints in softwood which is likely to form the basis of a future CEN Standard.

Regularly, joints are cut in widths of 225 or 250 mm but, before calling for joints in wider members, it is wise to check if a machine line is available with adequate capacity to cut such wide members, or members thicker than 75 mm.

USES OF FINGER JOINTS

Structural uses

Finger joints can be used for floor, roof and ceiling joists, studs, flanges of ply-web beams, glulam sections, members in trusses and trussed rafters, decking etc. In general, standards and codes call for at least 1 metre distance between finger joints in any one piece, but there is no reason to suppose that joints at closer centres weaken the piece. Such clauses are included mainly to anticipate anti-reaction of site staff etc. if joints were to occur repeatedly at close centres.

BS5268: Part 2: 1988 does not permit finger joints to be used where failure of one joint could lead to progressive or serious collapse. Thus, finger joints are permitted in floor joists but not in a single member trimmer beam supporting floor joists. If finger joints are necessary in a trimmer beam of solid timber, the trimmer can be made from two or more members equally strained.

The code of practice for trussed rafter roofs, BS5268: Part 3: 1985, places certain limitations on the use of finger joints under nail plates in the construction of trussed rafters made with thin members. There is no doubt that a nail plate does weaken the strength in bending and tension of a finger joint in thin timber if the teeth coincide with certain parts of the finger profile. This effect is catered for in BS5268: Part 2: 1988 by the introduction of coefficients from which the residual strength of a joint affected by the insertion of a nail plate can be calculated and can be compared with the required strength. In the majority of cases it will be found to be possible to use finger jointed timber in the manufacture of trussed rafters manufactured with metal plate gussets without incurring a design penalty.

Joinery uses

Finger joints are used in windows, door frames, skirtings etc. Finger jointing is now so generally accepted that the accent in codes and standards is to permit finger joints unless the specification excludes them. The correct profile must be used. Where clear finishes are to be used, if the timber pieces being joined are of similar colour and grain, short joints can be relatively unobtrusive. However, if pieces of different colour are joined, the colour change will be noticeable at the joint.

Cases have been reported where finger joints can be seen under an opaque paint finish. Certainly if the finish is not completely opaque and the profile is cut to appear on the exposed face (rather than at right angles to it), the regular shape of the joint is likely to be more noticed than irregular characteristics such as knots. With a fully opaque finish, however, and a cut (rather than a die-impressed) joint with no gaps at the tips, and correct moulding of sections, the joints should not be too obtrusive.

Some standards require finger joints in joinery to be a certain distance apart but this is purely a cosmetic requirement. In fact, in certain special cases, a manufacturer may decide to introduce finger joints at very close centres to add to the stability of a section.

QUALITY CONTROL

In the UK there is no mandatory requirement for manufacturers to participate in quality assurance schemes although there are voluntary schemes, and BS5291: 1984 and BS1186: Part 2: 1987 detail quality control requirements for structural and joinery finger joints respectively. To ensure reliable joints, manufacturers must exercise routine quality control on at least the following items:

- The moisture content of the timber.
- The temperature of the timber and air during curing of the adhesive. Pre-heating of the timber, or accelerated curing may be employed.
- Storage, mixing and application of the adhesive.
- Maintenance and operation of the machinery and cutters.
- Limiting the time between cutting and assembly of the joints.
- Controlling the end pressure at time of assembly of the joints.
- Checking the quality of the timber in and close to the joints.

In addition, it is essential for structural uses, and desirable for certain joinery applications, for the manufacturer to carry out routine tests to destruction on sample pieces. Note that, if finger jointed timber is passed through a stress grading machine, this will not give an adequate test of the strength of the finger joints, although it will give a form of proof loading.

ADHESIVES AND GLUING

The adhesive should be chosen to suit the durability requirement for the service conditions (see Tables B9.1 and B9.2 and Chapter B9 on glued joints). In addition, if a structural component such as a floor joist, rafter etc. is likely to be fully exposed to weather for some time during transport, storage and erection, it is preferable to glue the finger joints with a weather-proof and boil-proof adhesive (WBP) even if the component will be covered when in service. The trussed rafter code BS5268: Part 3: 1985 permits only WBP adhesives to be used in finger joints in trussed rafter material. Although BS5291: 1984 on structural finger jointing does not insist on the use of WBP adhesives, they are strongly favoured for structural applications.

Manufacturers produce adhesives specially modified for use on finger joints to prevent 'glue-starvation' (i.e. adhesive being forced into end grain) when the high end assembly pressure is applied. The modification usually takes the form of an additive such as stone-flour.

Adhesives must be compatible with preservatives etc. If a flame retardant containing ammonia is to be used, it must not be applied before finger jointing, but it may be possible to apply it after the adhesive has fully cured (i.e. probably seven or more days after the jointing process). The manufacturer of the adhesive should be consulted.

PROCESSING

If a planed or moulded section is to be produced, it is essential to finger joint before processing so that the pieces joined are in line after processing.

If a finger jointed member, such as a floor joist, is not processed after jointing, a certain amount of adhesive 'squeeze-out' will be visible and the pieces joined may not be exactly in line.

C1 NORDIC FORESTS AND SAWMILLING PRODUCTION OF REDWOOD AND WHITEWOOD

INTRODUCTION

The principal trees in the Nordic forests are pine (*Pinus sylvestris* L.) from which European Redwood is sawn, and spruce (*Picea abies* (L.) Karst) from which European Whitewood is sawn. In addition, there is birch, beech, aspen, oak and other deciduous trees, and experimental stands of evergreen trees not normally associated with the Nordic area. It is worth noting that, unlike mid-European spruce, Nordic spruce stands do not include Silver Fir (*Abies alba* Mill.).

Sweden and Finland are the major Nordic producers and exporters of timber. The sawmilling industries concentrate on the production of European Redwood and European Whitewood, but some of the beech etc. is used for furniture, and Finland uses birch in the manufacture of plywood. Norway is largely self-sufficient in the production of European Redwood and European Whitewood. Some Norwegian Whitewood is exported, mainly to the UK, but there is an import from Sweden and Finland. Denmark and Iceland are not regarded as major producers or exporters of European Redwood or European Whitewood, although timber windows, for example, are exported by Denmark to the UK. Denmark and Iceland are not mentioned further in these notes.

THE FORESTS

Sweden

Sweden is a long narrow country some 1600 kilometres long. It extends from latitude 55° North to latitude 69° North which is well into the Arctic Circle. The land area is 41 million hectares and, in 1980, some 24 million hectares were covered by productive forest. The forest area is gradually being increased as farm land in the north is given over to forestry, and as marsh land is drained for forestry.

Forestry and re-afforestation have been controlled strictly for over 200 years to ensure that felling does not exceed the annual increment of growth which, in 1980, was in the order of 82 million forest cubic metres over bark, 66 million of which was conifer. In 1980, the standing volume of pine was 970 million forest cubic metres over bark. The standing volume of spruce was 1160 million forest cubic metres over bark.

Finland

Finland extends some 1100 kilometres from latitude 60° North to latitude 70° North, therefore one third of the length is within the Arctic Circle. The land area is nearly 31 million hectares and, in 1980, some 20 million hectares were covered by productive forest. The forest area is being increased as marsh land is drained for forestry.

Forestry and re-afforestation are strictly controlled to ensure that felling does not exceed the annual increment of growth which, in 1980, was in the order of 63 million forest cubic metres over bark, 49 million of which was conifer. In 1980, the standing volume of pine was 709 million forest cubic metres over bark. The standing volume of spruce was 607 million forest cubic metres over bark.

Norway

Norway is a narrow and mountainous country extending from latitude 58° North to latitude 71° North. The land area is 37 million hectares (i.e. more than Finland) but, due to the terrain, less than 20% is covered by productive forest, the estimate in 1980 being 6.5 million hectares. The forests were over-exploited in the nineteenth century but, since then, a forestry policy has ensured that felling does not exceed the annual increment of growth.

In 1980, the annual increment of growth was in the order of 18 million forest cubic metres over bark (14 million of which was conifer). In 1980, the standing volume of pine was 154 million forest cubic metres over bark. The standing volume of spruce was 260 million forest cubic metres over bark. It is expected that the annual increment of growth will increase to 19 million forest cubic metres over bark.

Nordic Forestry

Redwood and Whitewood are spread fairly uniformly over the area of

Sweden, Finland and Norway. The best Redwood for joinery is considered to come mainly from trees which grow above latitude 58° North. The best Whitewood for quality uses comes from areas of Sweden and Finland known to specialist buyers. Both Redwood and Whitewood have similar strength from any area within Sweden, Finland or Norway.

Re-planting is by selected seedlings, or by natural regeneration in areas where the previous stand was of high quality. Seedlings are cultivated in nurseries from seeds from selected trees. Where commercially possible, stands of trees are thinned twice in the growing cycle, and pruning of branches takes place once or twice. Research and silviculture are aimed at continuing improvements in quality and uniformity. By the mid-1960s, forest inventories showed that over 40% of the pine and spruce trees had a diameter at breast height (dbh—the usual and convenient height for measuring tree diameter) of 250 mm or more. In general, the spruce trees have a somewhat larger diameter than the pine trees. As a general rule, one can estimate that the outer 50 mm is sapwood except in over-mature trees in which there is less sapwood.

In the extreme south of Sweden, trees grow for about 260 days per year, whilst in the north of all three countries, the weather is such that trees grow for only about 125 days per year. For the Nordic sawmilling industry, ideally trees should have a diameter at breast height of 300–400 mm or more. In the south, trees felled for sawn timber are 60–80 years old or older and, on average, have 6–8 or more growth rings per 25 mm. In the north, trees felled for sawn timber are 100–120 years old or older, sometimes considerably older, and the growth rings are much closer. At these ages, the trees will have reached a height of 25–30 m.

Although some felled trees are transported to a sawmill as full tree lengths, it is normal to cut felled and de-branched trees into logs in the forest. Normally logs vary in length from 3.0–5.8 m. Some are cut as short as 2.5 m, some in excess of 6.0 m. A modern development is to use a de-limbing machine in the forest which also cuts logs to the exact lengths required for final products. Most Nordic sawmills are designed to handle logs only up to 5.8–6.0 m in length.

In 1980, the forests yielded sufficient logs for the production of the volume of sawn timber listed below, as well as logs for pulp, paper, wood based panel products and fuel:

Sweden	11 000 000 m^3 of sawn timber;
Finland	9 000 000 m^3 of sawn timber;
Norway	2 100 000 m^3 of sawn timber.

The export to the UK in 1986 and 1987 in cubic metres and as a percentage of the import of softwood, was reported as:

	1986		1987	
Sweden	1 949 000 m^3	28%	1 856 000 m^3	23%
Finland	968 000 m^3	14%	1 059 000 m^3	13%
Norway	88 000 m^3	1%	84 000 m^3	1%

In 1973, the export from Sweden to the UK was as high as 3.19 million m^3, from Finland 2.03 million m^3, and from Norway 0.28 million m^3, but nowadays each Nordic country is involved more with producing specially dried, graded etc. timber rather than bulk supply. The capacity of the Swedish and Finnish sawmilling industry has been reduced accordingly, with the accent on using only higher quality logs.

SAWMILLING

Logs are transported to sawmills mainly by lorries. Logs are sorted by species and often by diameter, length and quality. The modern preference is to store logs on land rather than in water (to avoid the chance of areas of high-porosity in the sawn timber, particularly in sawn Redwood for joinery), sprayed with water as necessary to prevent premature drying-out and splitting of the surface of the logs. Mills which still use ponds to store and sort logs aim to limit the period of storage, particularly during warm weather.

Sweden

By 1984, the number of sawmills in Sweden had been reduced to 2500. Of these, seven have a capacity of over 150 000 m^3; 550 have a capacity of 10 000 m^3 or more. The capacity of the Swedish sawmilling industry has been reduced in recent years from about 14 million m^3 to about 11 million m^3, but with an increasing accent on the production of specialities rather than purely bulk production. Most significant sawmilling companies are members of the Swedish Wood Exporters' Association.

Finland

The number of sawmills in Finland is reported (in 1987) as being over 5200 but many have a capacity of less than 500 m^3. The number of significance is around 160, which account for 90% of the total capacity. Thirty have a capacity of over 100 000 m^3. The capacity of the industry has been reduced in recent years to below 8 million m^3. Most of the largest sawmilling companies are members of the Finnish Sawmill Owners' Association (to be renamed the Finntimber Association). Most of the other significant sawmilling companies are members of the Association of Finnish Sawmills.

Norway

The number of sawmills in Norway of significance is fewer than 400. The capacity of the industry is in the order of 2.1–2.3 million m^3.

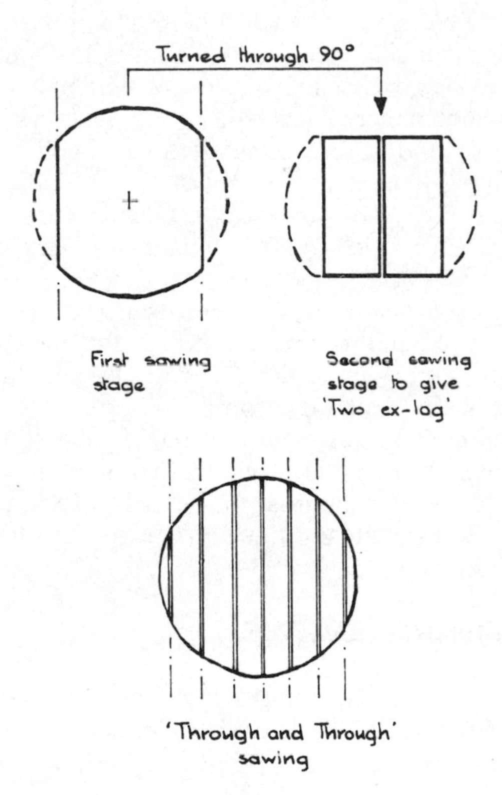

Fig. C1.1 Nordic sawmilling procedures

Nordic sawmilling procedures

Generally mills are highly mechanized and automated, making use of modern computer, closed-circuit TV and laser techniques to measure, control and record, particularly to determine the most efficient way to convert individual logs.

All Nordic mills operate with logs entering the mill end-to-end at intervals of a few seconds. After de-barking, the first saw position normally produces a 'centre cut', which is then turned through 90° before the second sawing position, where the centre is sawn 'two ex log' or 'three ex log' (see Fig. C1.1), or even into four pieces (not necessarily all the same thickness). Traditionally the 'frame saw' was used at both these sawing positions and is still used to a certain extent. However, many mills now use band saws or circular saws. One reason is that it is possible to adjust the position of each band saw or circular saw almost instantly by remote control, whereas to break down and re-set a frame saw is a manual operation which takes 15–20 minutes. Also, it is possible to replace a cassette band saw or a circular saw within a few minutes for maintenance. Each type of saw gives a different type of sawn surface. By and large, a circular saw or band saw gives a finer surface than a conventional frame saw.

Pieces are normally sawn oversize to take account of shrinkage on drying so that, when dried to 20% moisture content, they are full to size.

Parts of the log from outside the 'centre cut' (i.e. the side boards) are mostly sawn, edged and trimmed, although some mills chip them.

As a variation on the traditional sawing pattern, and particularly when dealing with small logs, some mills pass the log through a chipping machine to yield a square or rectangular section (or a more complex shape) which is then cut into two or more pieces. Such an operation yields no side boards. As a further variation, some mills practice types of 'through and through' sawing (see Fig. C1.1).

SAWN SIZES AND LENGTHS

Swedish, Finnish and Norwegian standards for sawn timber are similar, and similar to BS4471: 1987 for sizes of sawn and processed softwood.

The Swedish standard is SIS 23 27 11 *The Swedish standard for sawn and resawn timber*. The sawn sizes in it are reproduced in Table C1.1. The Finnish standard is SFS 2511 *Sawn softwood, dimensions*. The sawn sizes in it are reproduced in Table C1.2. The Norwegian standard is NS3042 *Sawn timber. Dimensions*. The sawn sizes in it are reproduced in Table C1.3. In addition, thicknesses of 47 mm are cut, and many mills are prepared to saw special sizes (even a Nordic version of the

Table C1.1 Sawn sizes from the Swedish Standard SIS 23 27 11: 1970 *The Swedish Standard for sawn and resawn timber*

Thickness (mm)	Width (mm)							
	75	100	115	125	150	175	200	225
16	**	**		*				
19	**	***	*	**				
22		***	*	**	**			
25		***	*	***	**	*		
32		*	*	*	*	*	*	
38		**	*	***	**	*	*	*
50		***	*	***	***	***	***	**
63		**	*	**	***	***	*	*
75		*		*	**	**	***	***

*** 'Preferred sizes'.
** 'Second choice sizes'.
* 'Dimensions to be used only in exceptional cases'.

Table C1.2 Sawn sizes from the Finnish Standard SFS 2511: 1982 *Sawn softwood dimensions*

Thickness (mm)	Width (mm)							
	75	100	115	125	150	175	200	225
16		*						
19	**	***	*	***				
22	*	***	*	***	*	*	*	
25		***	*	***	**	*	*	
32	*	*	*	*	*	*		
38		**	*	**	**	*		
44		*	*	*	*			
50	*	***	*	***	***	***	***	*
63					***	***		
75					***	***	***	**
100		***						
125				***				
150					*			

*** 'Sizes to be recommended primarily'.
** 'Sizes to be recommended secondarily'.
* 'Other sizes'.

Canadian 'surfaced' CLS sizes with rounded corners—see Chapter C5). When sawing, the aim is to eliminate minus tolerance when the timber is dried to 20% moisture content. Minus tolerance is permitted on only 10% of pieces and is limited to:

1 mm on sizes up to 99 mm, and
2 mm on sizes of 100 mm and over.

Plus tolerance is not limited.

Timber from Sweden and Finland is normally sold in length

Table C1.3 Sawn sizes from the Norwegian Standard NS3042: 1970 *Sawn timber. Dimensions*

Thickness (mm)	Width (mm)												
	25	38	50	63	75	100	115	125	140	150	175	200	225
13	***	**											
16		**				**		**					
19		**				***		**		**			
22						**		**		**			
25		**	***			***		**		**			
32			**		**	**		**		**			
38					***	***		**		***		***	
44					**	**				**		**	
50			**		***	***	**	**		***	**	***	**
63						**		**		**		***	
75					**	**				***	**	***	**
100						***							

*** 'Recommended dimensions that ought to be used'.
** 'Other dimensions that can be used'.

increments of 0.3 m (as BS4471: 1987, see Chapter B1) but Norwegian mills often cut to increments of 0.1 m. No minus tolerance is permitted. Overlength is limited to 50 mm. The maximum lengths normally available are 5.4 m or 5.7 m unless pieces are finger jointed in which case length is limited only by handling and transportation considerations. Finger jointed lengths of 10 m and 12 m of stress graded sizes are exported to the UK. By special arrangement, timber can be cut to specific lengths within close tolerances.

DRYING

Unless specifically requested to sell un-dried timber, all Nordic timber sold to the UK is dried. Well over 90% is kiln dried, the remainder being air dried. The majority of timber is dried in a progressive kiln (see Chapter B2), but there is considerable investment in chamber kilns.

To prevent sap stain occurring in Redwood, until the 1970s it was normal to dip or spray the timber with an anti-stain formulation, usually based on pentachlorophenol (p.c.p.) prior to kilning. For environmental reasons, p.c.p. is now banned in Sweden and Finland and is not used in Norway. Nowadays, many Nordic mills do not use anti-stain treatments, but reduce the chance of sap stain occurring by commencing kilning within a few days of sawing, and by slightly reducing the upper limit of the moisture content spread in kilned timber. Some mills still apply an anti-stain during the warmer months of the year (e.g. from May until September), but use a formulation cleared for environmental safety.

To reduce the use of fuel oil etc., much of the power to heat kilns comes from the residue from logs.

In recent years, in Sweden and Finland, there has been a great deal of investment in chamber kilns (see Chapter B2). By using a chamber kiln, sawmills are able to offer to dry timber to moisture contents appropriate for joinery, pine furniture etc. It is quite common, by special arrangement, to dry timber to average levels as low as 10/12%, or even to an average of 8%, with only a small spread of moisture content.

SAWMILL GRADES

After the timber has been dried it is graded. In Sweden and Finland, most is graded respectively to the 'Guiding principles for grading of Swedish sawn timber', which is presented in English in a publication produced by The Association of Swedish Sawmillmen and published by *The Swedish Timber and Wood Pulp Journal*, and the Finnish 'Instructions for the grading of export timber', available in English from the Association of Finnish Sawmillmen, Fabianinkatu 29C, Helsinki, both of which are essentially similar. Much of the export of Norwegian timber is to similar rules, although Norwegian rules for the domestic market are less rigid in certain aspects (e.g. on wane).

It should be noted that the Swedish and Finnish books describe 'guiding principles' rather than firm rules. Many mills operate their own (usually tighter) variations, hence the phrase 'usual bracking' (i.e. the usual quality to be expected from this particular mill). These are not end use grades and hence are of limited interest to specifiers of building timber, although it is important for specifiers to be aware of the background to these guiding principles, if only not to be confused by them.

The guiding principles describe six 'basic qualities':

I, II, III, IV, V and VI,

with I being the top quality. The qualities are based mainly on the *better* face and two edges, therefore the term 'knot free timber' to a sawmiller means free from knots on one face and two edges, not free from knots all round. (Even so, normally only small or short pieces of Nordic timber are available 'knot free'.)

Particularly for the export markets, it is normal for mills to group together the top four basic qualities. Because they are not separately sorted they are given the title 'unsorted' (not 'unassorted' as one hears occasionally). This is a simplified explanation of why the best timber is given a rather uncomplimentary sounding name. There are no firm rules on the percentage of the I–IV qualities which should occur in unsorted.

The 'fifth quality' (V) is normally sold separately, as is the 'sixth quality' (VI), but there are variations such as 'fifth and better'. Sometimes a mill will sell 'saw-falling quality', which can mean either 'fifth and better' or can permit a small percentage of 'sixth' to be included. If there is doubt when dealing with a production for the first time, it is better to ask what is intended. Although it is normal for Redwood and Whitewood to be sold separately, sometimes the species are mixed in the lower qualities, or in stress graded timber.

Phrases such as 'Upper Gulf Swedish', 'Group 2 Finnish' etc., are used by, and may be meaningful to a specialist buyer or seller, particularly if the buyer has knowledge of the particular mill. They are not precise, however, in the context of an end use specification for building timber, particularly if the timber is to be resawn. Normally such phrases should not be included in any end use specification. However, because a specifier may come across them, it can be valuable for specifiers to understand what is intended. The following notes are given with that in mind, not with the intention that such phrases should be used in end use specifications for building timber. Traditionally it was considered that any term which indicated a more northern production implied better quality for joinery timber, but that is not necessarily the case, although certainly it is most likely to indicate the closest growth rings.

It is worth mentioning that 'The Gulf' referred to is the Gulf of Bothnia not the whole length of Sweden as is sometimes imagined, therefore there are only a limited number of Upper Gulf mills (although there are many more 'North Swedish Mills'). Note also that the phrase, when used by specialist buyers, refers only to Swedish mills, not to Finnish mills, even though shores of 'The Gulf' are in Finland. Specifiers may come across the phrases 'Group 1 Finnish', 'Group 2 Finnish' etc. When used by a specialist buyer or seller, the reference to a 'Group' is for Finnish mills, not Swedish mills. Group 1 indicates the highest quality. It should be realized, however, that there is no official inspection before a mill qualifies for a particular Group. The choice is left to each mill to decide, and can even vary from year to year depending on the quality of the logs which will be available and the marketing plans for the mill. A mill will often decide to offer its Redwood as fitting one Group with its Whitewood fitting another Group. There are four 'Redwood Groups' and three 'Whitewood Groups'. There is no firm geographical zoning for Group qualification, no written quality rules, and no published list of the mills which have decided to market their timber as fitting a particular image. Such terms are best avoided in end use specifications.

Occasionally, buyers can be confused when they find out that a production described as 'North Finnish' is shipped from a port in the south on the Gulf of Finland. This is because there is a main north/south rail link which is used sometimes, particularly by companies situated in the north-east if parts of The Baltic are frozen, instead of the preferred method of shipping from a North Finnish port.

Occasionally, buyers may come across traditional sawmilling terms such as 'schaal-boards', Halmstad Baulks etc., with which they are not familiar. In such cases it is prudent to check the exact meaning with the importer or agent. Thin schaal-boards are normally largely knot-free on the outside face but a considerable amount of wane is permitted. Halmstad Baulks vary from 75 × 75 mm to 200 × 200 mm or larger and are permitted to have a considerable amount of wane. In recent years, new terms such as 'green-split', 'centre-free' etc. have been given to timber specially selected for joinery. When encountering a new term for the first time, a clear definition should be requested.

SHIPPING MARKS

After grading, one end (or both ends) of every piece of timber is marked (normally in red) with a 'shipping mark' which identifies the sawmill (or grading company) and the quality of the timber. As a general rule, a 'Crown' in the mark of Swedish or Finnish timber normally indicates 'unsorted' quality. Cross-reference publications of shipping marks are: *Shipping marks on timber*, published by Benn Publications Ltd. (currently out of print), and *Handbook of the northern wood industries*, published by *The Swedish Timber and Wood Pulp Journal*.

PACKAGING AND WRAPPING

All Nordic export sawn timber is packaged, and a large percentage of packages are wrapped in several different ways. The normal package fits within the quarter-container module (i.e. width close to but not in excess of 1.2 m; height close to but not in excess of 1.0 m). Each package contains timber of the same size and quality. There are variations, particularly when specially graded or dried timber is being supplied.

Some packages are 'truck-bundled' (i.e. containing timber of varying lengths), some are length packaged or contain timber of two or more different lengths.

SPECIAL GRADING, MOULDING ETC.

As well as special drying, Nordic mills carry out a significant amount of special grading, planing, moulding etc., particularly for the UK market, and normally in accordance with British Standards, for example:

- Timber stress graded visually or by machine to BS4978.
- Planed trussed rafter material.
- Floor joists.
- Tiling battens, preserved if required.
- Special joinery qualities.
- Long lengths/ pre-cut lengths/ finger jointed timber.
- Tongued and grooved mouldings, shiplap, battens etc.
- Shrink-wrapping of products.
- Door linings or door frame sets.
- Laminated and edge jointed sections.

Etc.

C2

RUSSIAN FORESTS AND SAWMILLING PRODUCTION
OF REDWOOD AND WHITEWOOD

THE FORESTS

Russia is a vast country. It extends from latitude 38° North to latitude 77° North which is well inside the Arctic Circle. Most is above latitude 45°. The land area is over 2100 million hectares. There are several estimates of the extent of forests which can vary depending on the definition of forest. A recent estimate gave 790 million hectares as the area of productive forest. Another estimate, quoted by the Russian Wood Agency, gives 930 million hectares if one includes 'other wooded land'. About a quarter of the forest is in 'European Russia', three quarters in 'Asiatic Russia'.

The latest estimate of the standing volume of trees is 84 200 million forest cubic metres over bark, but much is inaccessible due to the great distances involved and the severe climate. It is likely that access to the eastern forests will be improved but a percentage of the trees are likely to be over-mature.

Of the 790 million hectares of productive forest, 37% is estimated to be larch, 21% pine and 11% spruce, the remainder being broadleaved trees of which over half is birch. The volume of conifer (including larch) is estimated to be 67 000 million forest cubic metres over bark. Probably pine and spruce account for slightly more than half of this volume.

The forests are owned by the State, which also controls the amount of sawn timber that is exported. Russian sources state that during 1971–5, 10.4 million hectares were re-afforested, of which more than 5 million were sown or planted. The 10th 5-year plan (1976–80) provided for between 10 and 11 million hectares of re-afforestation.

The two principal coniferous trees for sawn timber are pine (*Pinus sylvestris* L.) from which European Redwood is sawn, and spruce (*Picea abies* (L.) Karst) from which European Whitewood is sawn. In addition, there is a considerable amount of larch. The main hardwoods are birch, aspen, oak and beech. Birch is used for plywood. It is estimated that the current production of European Redwood and European Whitewood by the Russian sawmilling industry (103 million m^3) is about one third of the World's production of sawn softwood. At least 90–92% of the production is used on the domestic market but the export can still be in excess of 8 million m^3 of sawn timber, although in recent years it has been nearer 7 million m^3. The timber for export is produced by specialist mills which concentrate on export markets. The export to the UK in 1986 was 1 062 000 m^3 which was 15% of the import and, in 1987, was 1 286 000 m^3 which was 16% of the import. Of the Russian export to the UK, about 70% is European Redwood.

The growth pattern (number of growing days per year) of the pine and spruce trees is similar to that for similar latitudes of Sweden and Finland, but a percentage of the trees have a larger diameter before they are felled. The finest Redwood is traditionally considered to be shipped from the Kara Sea area with the best Whitewood being shipped from the White Sea area.

SAWMILLING AND SAWN TIMBER

There are a considerable number of sawmills of significant size in Russia (over 1200). The sawmilling industry concentrates on bulk production. The sizes offered are shown in Table C2.1 as taken from the 1989 'Russian Schedule'. The largest mill complexes have a capacity of up to 600 000 m^3 (e.g. Mill 16/17 in Archangel).

A modern icebreaker fleet has extended the 'shipping season' which now operates from February to December.

The industry has adopted the policy of aiming to eliminate minus tolerance in sawn timber when dried to 20% moisture content. Lengths are in increments of 0.3 m up to 6.0 m. Specifiers should consider Russian timber as complying with the tolerances of BS4471: 1987. Most lengths are '2.7 m and upwards' but some 1.5 m–2.4 m lengths are offered.

The frame saw is still very popular although band saws and circular saws are also used. Chipper saws are used for smaller diameter trees. The sawmilling procedures are similar to those described in Chapter C1 for the Nordic area. Many mills have kilns but some still use air drying. Where kilns are installed, the moisture content of the dried timber is around 18% average as described for Swedish and Finnish mills. Where air drying is used, the average moisture content is likely to be slightly higher. The complexes are being provided with progressive kilns. Normally, Russian timber is dried before export to the UK. Generally, both the Redwood and Whitewood are currently treated with anti-stain.

Table C2.1 Sizes from the 'Russian Schedule' of 1989

Sizes shipped from the Kara Sea area

Thickness (mm)	Width (mm)										
	100	115	125	140	150	160	175	200	225	250	275
19	×		×		×		×	×	×		×
25	×		×		×		×	×	×		×
32	×		×		×		×	×	×		×
38	×		×		×		×	×	×		×
50	×		×		×		×	×	×		×
63	×		×		×		×	×	×		
75	×		×		×		×	×	×		

Sizes shipped from areas/towns other than the Kara Sea area

Thickness (mm)	Width (mm)											
	100	115	125	140	150	160	175	200	225	250	275	300
16	×	×	×	×	×	×	×	×	×	×	×	
19	×	×	×	×	×	×	×	×	×	×	×	
22	×	×	×	×	×	×	×	×	×	×	×	
25	×	×	×	×	×	×	×	×	×	×	×	
32	×	×	×	×	×	×	×	×	×	×	×	
38	×	×	×	×	×	×	×	×	×	×	×	
44	×	×	×	×	×	×	×	×	×	×	×	
47	×	×	×	×	×	×	×	×	×	×	×	
50	×	×	×	×	×	×	×	×	×	×	×	
63	×	×	×	×	×	×	×	×	×	×	×	
75	×	×	×	×	×	×	×	×	×	×	×	

MILL GRADING

The grading rules are not published outside Russia but the timber trade consider that mills adhere to the rules. In 1985, the 'Leningrad bracking rules' and the 'White Sea bracking rules' were replaced by the 'Northern bracking rules'.

The rules cover:

'unsorted',
'fourth quality' and
'fifth quality'.

These are not end use grades.

It is convenient to compare the basic grades to the Swedish and Finnish rules except that the 'unsorted' (similar to Nordic 'unsorted') consists of Russian I, II and III basic qualities, whilst the 'fourth' (IV) quality is similar to Nordic 'fifth' quality, and the 'fifth' (V) quality is similar to Nordic 'sixth' quality.

The grades are indicated by 'shipping marks' on both ends of each piece. They are given in *Shipping marks on timber*, published by Benn Publications Ltd. (currently out of print), and *Handbook of the northern wood industries*, published by the *Swedish Timber and Wood Pulp Journal*.

Russian shipping marks identify the area or town from which the timber is shipped and, in addition, there is one mark to identify the 'unsorted', one for 'fourth quality', and one for 'fifth quality'. All marks begin with an E which stands for Exportles (the Russian agency for the export of timber). Two stars indicate 'unsorted'. One star indicates 'fourth quality'. A line indicates 'fifth quality'. The marks used to be put on with a 'hammer' but now are in red on both ends of each piece.

The marks indicate either one of three areas:

Leningrad, Petchora and the Kara Sea area,

or one of ten towns:

Archangel, Onega, Mesane, Oumba, Kem, Pudoj, Medvezhjegorsk, Petrozavodsk, Segezha and Belomorsk.

PACKAGING AND WRAPPING

All Russian export sawn timber is packaged to the quarter-container module (i.e. width close to but not in excess of 1.2 m, height close to but not in excess of 1.0 m). Each package contains timber of the same size and quality. Packages are banded. The packages produced by large complexes are pressed before banding and protected by waterproof paper.

Although most Russian packages contain timber of two or more lengths (usually in consecutive 0.3 m increments), most Archangel Whitewood is length packaged and a percentage of the Redwood is length packaged. Although more Russian Whitewood for export is likely to be length packaged in future as more use is made of the export mill complexes, it is considered likely that most of the Redwood will continue to be sold with two or more lengths in each package.

C3 POLISH FORESTS AND SAWMILLING PRODUCTION OF REDWOOD AND WHITEWOOD

THE FORESTS

Poland is situated between latitudes 49° North and 55° North on the Lower Baltic. The country has a land area of 39 million hectares. An area of 8.32 million hectares is covered by productive forests of which 6.56 million hectares are coniferous forests. The growing stock of conifer is some 830 million forest cubic metres over bark.

Ninety-two per cent of the coniferous forest area is pine (*Pinus sylvestris* L.) from which European Redwood is sawn. Eight per cent is spruce (*Picea abies* (L.) Karst) from which European Whitewood is sawn. The Whitewood may also include some Silver Fir (*Abies alba* Mill.). Although the forests lie further south than those of Sweden, Finland and Russia, the sawn timber is considered to have the same strength when stress graded to the UK visual stress grading rules as timber from the Nordic area.

The trees grow for more than 260 days per year and have an average of around 6 to 8 growth rings per 25 mm. Currently the production of coniferous sawn timber is about 4.5 million m^3 of which almost 90% is used on the domestic market. Currently the UK takes about 40% of the export from Poland. In 1970 the export to the UK was 489 000 m^3 but, in 1986, the export to the UK was 189 000 m^3 which was under 3% of the import. The corresponding figures for 1987 were 200 000 m^3 and about 2.5%.

SAWMILLING AND SAWN TIMBER

The sawmilling procedures are generally similar to those described in Chapter C1 for the Nordic area but on a smaller scale. Since the early 1970s there has been considerable investment in kilns. Although most kilns are used for timber for the domestic market, there is some export of kilned material; however, most timber exported is air dried. There are about 600 sawmills of which about 350 are 'on the export list', including for hardwoods. The timber is sawn oversize so that it is the full basic sawn size when dried to 20% moisture content. Tolerances on sizes can be considered to be as in BS4471: 1987 (see Chapter B1).

The sizes and lengths available can be assumed to be the same as those in the Swedish standard (see Table C1.1) plus thicknesses of 47 mm.

Where the timber is air dried or kiln dried, the aim is to dry most pieces to 20/22% moisture content or lower (20% when kiln dried), but certain pieces will have parts at slightly higher levels.

Although the Redwood is not considered to be as suitable for joinery as Redwood from the Nordic area, it accounts for about 75% of the export, including the export to the UK where it is used mainly as a construction timber.

The majority of the export timber is packaged in widths between 1.0 m and 1.125 m, and to a height around 0.6 m, although some packages are 0.6 m × 0.6 m. Most timber is packaged to length but, as an alternative, some packages contain timber with a maximum of four consecutive length increments of 0.3 m.

All packages are wrapped with plastic or waterproof paper placed under the top layer of timber.

It is not normal for the timber to be treated with anti-stain.

MILL GRADING AND MARKING

The mill grading system (quality grading) is based on a set of traditional Swedish rules (the Härnösand District Rules) for 'unsorted', 'fifth' and 'sixth' quality (see Chapter C1). In general, growth rings will be a little wider than those in Nordic timber.

All mills use the same series of shipping marks to indicate the quality of the timber. These are variations on the letters L, P and an eagle. The marks are given in *Shipping marks on timber*, published by Benn Publications Ltd. (currently out of print), and *Handbook of the northern wood industries*, published by the *Swedish Timber and Wood Pulp Journal*. All export is arranged by PAGED, which is the Foreign Trade Enterprise, Warsaw.

Normally, the export to the UK is mainly 'fifth quality' Redwood and 'saw-falling' Whitewood. (Polish saw-falling consists of unsorted plus fifth quality.) Also, Poland is authorized to stress grade visually to satisfy the 'General Structural' and 'Special Structural' rules of BS4978: 1988, and has installed a machine which stress grades to BS4978: 1988 within the BSI 'Kitemark' scheme.

C4 CZECHOSLOVAKIAN FORESTS AND SAWMILLING PRODUCTION OF WHITEWOOD

THE FORESTS

Czechoslovakia is orientated mainly east to west, and most of the country is situated between latitudes 48° North and 51° North. The land area is 13 million hectares. Some 4.1 million hectares are covered by productive coniferous forests, mostly in the west. The growing stock of conifer is over 450 million forest cubic metres over bark. The principal conifer is spruce (*Picea abies* (L.) Karst.) with some Silver Fir (*Abies alba* Mill.). European Whitewood is sawn from both species. Although the forests lie further south than those of Sweden, Finland and Russia, Czechoslovakian Whitewood is considered to have the same strength when visually stress graded as Whitewood from the Nordic area.

The trees grow for more than 260 days per year and have an average around 6 to 8 growth rings per 25 mm. Currently the forests are reported to be able to yield around 4.2 million m^3 of sawn softwood per year of which nearly 75% is used on the domestic market. Around 1.2 million m^3 is available for export. Some of the trees have a somewhat larger diameter than Nordic trees, therefore it is possible to buy widths of up to 300 mm in lengths up to 6.0 m. The export to the UK in 1987 was 238 000 m^3 which represented 3% of the total import. The corresponding figures for 1988 were 225 000 m^3 and a little under 3%.

SAWMILLING AND SAWN TIMBER

The sawmilling procedures are generally similar to those described in Chapter C1 for the Nordic area but not on the same scale. Normally it is 'unsorted' Whitewood which is exported to the UK. Since the mid-1970s there has been considerable investment in kilns and it is

Table C4.1 Czechoslovakian sawn sizes based on the 1989 Schedule

Thickness (mm)	Width (mm)								
	100	125	150	175	200	225	250	275	300
22	×	×	×	×	×				
32	×	×	×	×	×	×	×	×	×
36	×	×	×	×	×	×	×	×	×
38	×	×	×	×	×	×	×	×	×
44	×	×	×	×	×	×	×	×	×
47	×	×	×	×	×	×	×	×	×
50	×	×	×	×	×	×	×	×	×
60[1]	×	×	×	×	×	×	×	×	×
75	×	×	×	×	×	×	×	×	×

[1] Not 63 mm.

anticipated that, in the immediate future, about half of the export will be kilned. There are reported to be about 600 sawmills of which about 100 are involved with the export market. The total production is currently around 4.2 million m^3, having increased from 3.5 million m^3 in 1978. The timber is sawn oversize so that it is the full basic sawn size when dried to 20% moisture content.

The sawn sizes of unsorted Whitewood which are sold on the export market are tabulated in Table C4.1, based on the 1989 'Schedule of prices for Czechoslovakian whitewood offered by the Ligna Foreign Trade Corporation, Prague'. A popular size in the UK is 38 × 225 mm for scaffold boards. On the other hand, it is not normal for 60 mm thicknesses to be bought for the UK market.

Tolerances on sawn sizes can be considered to be as BS4471: 1987 (see Chapter B1).

Generally, lengths vary in 0.3 m increments, usually between 2.7 m and 6.0 m, the average being around 3.9 m–4.2 m. Special length mixes are available at a small extra cost.

Most timber is exported dried (kiln dried or air dried: KD/AD). Even goods not sold specifically as being KD/AD are required to be dried sufficiently to arrive 'bright at the port of discharge'.

Czechoslovakia has been approved to produce machine stress graded timber within the BSI Kitemark scheme.

Some packages are made up from random lengths (truck-bundled) whilst some are packaged to length. As an alternative, some packages can be arranged to include a maximum of three consecutive lengths (i.e. differing by 0.3 m between each length). Packages are usually of the quarter-container module size (i.e. width close to but not in excess of 1.2 m, height close to but not in excess of 1.0 m), although some packages are 0.6 m wide by 0.5 m high.

Packages supplied to the UK used to be sent by rail to a port in Western Europe (usually Rotterdam) and then forwarded by ship. However, in 1968, LIGNA (the Foreign Trade Corporation for Export and Import of Timber and Products of the Woodworking and Paper Industries) established a terminal at Stettin from which timber is now shipped after having been transported by rail through Poland.

Czechoslovakian Whitewood is not treated with anti-stain. At least half the volume is shipped with waterproof paper wrapping under the top layer of timber. No planed timber has been exported to the UK to date.

MILL GRADING

Shipping marks on timber by Benn Publications Ltd. (currently out of print) gives the mill grading as 'unsorted', 'mill run quality' and 'fourth quality', although the only grade exported to the UK is 'unsorted' Whitewood. The quality of this may be related to unsorted Whitewood from the south of Sweden, but slightly faster-grown.

All Czechoslovakian unsorted Whitewood is given the same shipping mark 'LIGNA' marked in red on the end of each piece. (LGA in red is the symbol for 'mill run', and LGA in green is the symbol for 'fourth' quality.)

C5 CANADIAN FORESTS AND SAWMILLING PRODUCTION

OF SPRUCE-PINE-FIR, HEMLOCK OR HEM-FIR, DOUGLAS FIR-LARCH AND WESTERN RED CEDAR

THE FORESTS

Canada extends from Latitude 42° North to within a few hundred miles of the North Pole. Most of the area is above latitude 49° North. The area is 997 million hectares. The most important forest areas are in British Columbia in the west, where it is estimated that over half of the area of 94.9 million hectares is covered by forest, mostly of conifer, with most of the timber coming from 28 million economically productive hectares (i.e. 29.4% of the total area). In addition, Eastern Canada (Ontario, Quebec, New Brunswick, Nova Scotia, Prince Edward Island and Newfoundland) contains over 100 million hectares of forest of which some 90% are of conifer.

The total area of forest in Canada is estimated at around 20% of the total land area. Although the area of British Columbia is less than 10% of the total land area of Canada, about 60% of Canada's sawn softwood is taken from British Columbia forests (coast and interior), and about 75% of Canada's export of softwood is from British Columbia. Overall, felling is in balance with the annual increment of growth although some species (e.g. Douglas Fir) are less available than in the past, and other species (e.g. the Spruce-Pine-Fir grouping) are now sold much more than in the past. In Canada, about 169 million cubic metres of 'roundwood' (i.e. mostly softwood but some hardwood) was cut in 1985. In 1987, the Canadian production of sawn softwood was 62 million cubic metres (60% from British Columbia).

BRITISH COLUMBIA

Several principal conifers are grown in British Columbia. They are:

1. Western Hemlock (*Tsuga heterophylla* (Raf.) Sarg.)
2. Amabilis Fir (*Abies amabilis* Dougl. ex Forb.)

   Western Hemlock grows to a height of 35–55 m and is felled when the diameter at breast height is 900–1200 mm. Amabilis Fir grows to a height of 25–35 m and is felled when the diameter is 600–900 mm. The species are similar and are mixed in parcels of sawn timber. When sold for joinery etc. the trade term used is 'Hemlock'. When sold for stress graded uses the trade term used is 'Hem-Fir'. Hemlock and Hem-Fir can contain Amabilis Fir.

3. White Spruce (*Picea glauca* (Moench) Voss. var albertiana (S.Brown) Sarg.)
4. Engelmann Spruce (*Picea engelmannii* (Parry) Engelm.)
5. Lodgepole Pine (*Pinus contorta* Dougl. var latifolia S.Watson)
6. Alpine Fir (*Abies lasiocarpa* (Hook.) Nutt.)

   These four species grow to a height of up to 30 m, and are felled when the diameter is up to 900 mm, with 600 mm being a more common maximum. They are sold mainly in a stress graded mixed grouping known as 'Spruce-Pine-Fir'. (Also see 12 below.) There is no set percentage for each species in the mix, but the spruces are predominant, indeed a recent development is to separate and sell the spruces without the inclusion of pine or fir.

7. Douglas Fir (*Pseudotsuga menziesii* (Mirb.) Franco.)
8. Western Larch (*Larix occidentalis* Nutt.)

   These species grow to a height of 35–55 m and a diameter of 600–1800 mm. Although most of any consignment is Douglas Fir, some Larch may be included.

9. Western Red Cedar (*Thuja plicata* D.Don)

   This species grows to a height of 30–45 m and a diameter of 900–2400 mm. It is sold as a separate species.

In addition, from British Columbia, there are:

10. Sitka Spruce (*Picea sitchensis* (Bong.) Carr.)
11. Yellow Cedar (*Chamaecyparis nootkatensis* (D.Don) Spach.), which is really a cypress.

These two species have not normally been sold to the UK in recent years.

EASTERN CANADA

The principal species from the east of Canada are:

12. A mixed grouping of 'Spruce-Pine-Fir' which can contain:

    White Spruce (*Picea glauca* (Moench.) Voss.)
    Red Spruce (*Picea rubens* Sarg.)
    Black Spruce (*Picea mariana* (Mill.) Britt., Sterns. and Pogg.)
    Jack Pine (*Pinus banksiana* Lamb.)
    Balsam Fir (*Abies balsamea* (L.) Mill.)

13. White Pine (*Pinus strobus* L.)
14. Red Pine (*Pinus resinosa* Ait.).

As well as the species listed as Numbers 3,4,5,6 and 12, the UK structural design code also permits Ponderosa Pine (*Pinus ponderosa* Dougl.) from British Columbia to be included in a Spruce-Pine-Fir mix.

SAWMILLING

The trees are cut into logs in the forest. Logs can be up to 15 m long. They are transported to sawmills by water or road. There are some 350 sawmills in British Columbia out of a total of 1700 in Canada. The largest have a capacity approaching 500 000 m^3 per annum.

Logs are sorted and stored in water or on land at the sawmill. In the large coastal sawmills, the first sawing stage on a large log is carried out on a 'head rig'. This usually consists of a moving carriage and a large band saw. The log is clamped to the carriage which moves backwards and forwards in front of the operative (the 'head' sawyer) who is able to give individual attention to each log. The operative can rotate the log through 360° and is able to separate slabs of timber which are largely knot free from the centre part (cant) which includes knots. The cant passes onwards to edgers or gang saws. The outside slabs pass on to be sawn into sizes required for joinery or special uses.

In the sawmills in the interior, and in those which deal with smaller logs, the sequence of sawing is closer to the Nordic method described in Chapter C1. The first sawing stage is usually by band saws. Interior mills concentrate mainly on timber for construction purposes rather than joinery.

From larger trees, much of the outer part which is largely knot free is sold as 'Clears' or 'Factory Lumber' for conversion (re-sawing) by the final buyer. (See later notes in this chapter for a description of 'Clears' and 'Factory Lumber'.)

Side boards are edged (i.e. cut to width). Ends are trimmed to lengths in increments of 2 feet (or occasionally in increments of 1 feet).

Normally over half of the Canadian production is exported to the USA. Most of this is transported by rail or road. The export to Japan is about 6%, and the export to Europe (including the UK) is about 7%. Normally, to date, Canadian sawn or surfaced timber has been exported to the UK undried ('green'). The exception is 'Spruce-Pine-Fir' in surfaced CLS (Canadian Lumber Standards) sizes which is usually kiln dried to 19% moisture content or less. Discussions are taking place about drying of more Canadian timber. The drying of CLS may be by high temperature kilning in which schedules can be as short as 25–50 hours and temperatures can be as high as 105–115°C. Sections kilned in this way are not normally intended to be resawn. If deep sawn (i.e. parallel to the longer dimension), there can be a tendency to cup.

The timber which is sold 'green' is treated with an anti-stain. This used to be based on pentachlorophenol (p.c.p.), but due to environmental concern about this chemical, other anti-stains are now being used.

Other than for large sizes, Canadian timber is packaged normally 4 feet wide and 2 feet high. Large sizes are packaged to suit the order. Packages are square both ends. Kiln dried packages are wrapped.

SAWN TIMBER

The Canadian sawmilling industry in total has not yet adopted metric measure to any great extent, therefore most Canadian sawn sizes available in the UK are Imperial sizes with tolerances to the National Lumber Grades Authority (NLGA) 1987 Rules, although an increasing percentage of the sawn timber exported to the UK is of the metric sizes quoted in BS4471: 1987. Sometimes the sizes quoted are for when the timber is 'dry', sometimes when 'green'. Although the term 'nominal' in relation to sizes is no longer used in BS4471: 1987, it is still used often in the timber trade. It appears in the NLGA Standard Grading Rules for Canadian Lumber, in which Paragraph 44 on 'Actual Sizes' states:

> 'The use of "nominal" sizes in the language of these rules is for convenience and follows the practice of the industry. No inference should be drawn that the "nominal" sizes are actual sizes.'

When the term 'nominal' is used by the timber trade in relation to sawn sizes (rather than surfaced CLS sizes—see below), specifiers can regard it as being the size to which tolerances apply. It is normally also the size on which cost is based. Canadian sources tend to use the phrase 'rough sawn' in relation to sawn sizes, therefore this is the phrase used in the headings to Tables C5.3, 4, 5 and 6.

Use of the term 'nominal' size in relation to surfaced CLS (see later notes on CLS) is not a reference to the size to which tolerances apply, but is normally the size on which cost is based.

Sometimes Imperial sizes of sawn timber exported from Canada to the UK are 'sold-on' by a UK company as though being a metric size. For example $1\frac{7}{8}$ inch × 8 inch sizes are sometimes sold as being 47 × 200 mm. On the other hand, some Canadian mills saw a proportion of their production for the UK in metric sizes. Before using metric sizes in designs, care should be taken therefore to establish if the quoted size is when dry or when green, and if the tolerances on size are those of BS4471: 1987 or NLGA tolerances.

In considering what Canadian timber is available in the UK for various end uses, it is necessary to link the sizes which are available in the various species/grade combinations and to take account of the different tolerances relevant to each combination. This is outlined in the following pages.

Grading is in accordance with the 'National Grading Rules for Dimension Lumber' (NLGA: 1987) published by The National Lumber Grades Authority of Canada, or the 'Export R List Grading and Dressing Rules' adopted by the British Columbia Lumber Manufacturers' Association, or to BS4978. Stress graded timber has stamps on the face which include the mark of the agency which controls the grading. Shipping marks are given in *Shipping marks on timber*, published by Benn Publications Ltd. (currently out of print).

The total export of sawn and 'surfaced' (see below) timber to the UK in 1987, based on nominal sizes, was 2 332 000 m^3 which represented 29% of the total import. The figures for 1986 were 1 537 000 m^3 and 22%. (The large variation is generally accepted as being due to variations in exchange rates.)

STRESS GRADED TIMBER - SURFACED ON FOUR SIDES - CLS

Canadian stress graded timber, surfaced on four sides and with rounded corners, is referred to as CLS (Canadian Lumber Standards) sizes. The sizes normally available are given in Tables C5.1 and C5.2. Although such timber is occasionally available in Douglas Fir-Larch, CLS is available mainly in Spruce-Pine-Fir and Hem-Fir. The Spruce-Pine-Fir is kiln dried to 19% moisture content or less. The Hem-Fir is normally shipped 'green'. The 'surfacing' is similar to the type of

Table C5.1 Wider sizes of CLS

Nominal size (inches)	Basic size when dry (mm)
2 × 6	38 × 140
2 × 8	38 × 184
2 × 10	38 × 235
2 × 12	38 × 285

See text for tolerances which are related to the basic size.

'regularizing' explained in BS4471: 1987 (rather than planing as associated with joinery) but the timber is surfaced on both width and thickness. Corners are rounded not more than 3 mm.

The stress grades available in the UK for the wider sections of CLS as given in Table C5.1 used to be mainly a mix of 'No. 1 and No. 2 Structural Joist and Plank'. However, the UK structural design code BS5268: Part 2: 1988 now allocates the same stresses to No. 1 and No. 2. As a consequence, the higher stress grade of 'Select Structural Joist and Plank' is now more readily available and the mix of No. 1 and No. 2 is replaced simply by No. 2. Although Select Structural and No. 2 are given separate marks, they are permitted to be packaged together. Stresses for a 'No. 3 Structural Joist and Plank' grade are included in BS5268: Part 2: 1988 but this grade is not normally available in the UK. Machine stress graded CLS is now being supplied to the UK.

The tolerances on the basic sizes when dried to 19% moisture content are (according to BS4471: 1987 and the NLGA Rules): no minus tolerance, with plus tolerance unlimited. (The plus tolerance is usually small.)

Lengths are in 2 feet increments from 8 to 24 feet.

The stress grades which are relevant for the narrower CLS sections as given in Table C5.2 are 'Select Structural Light Framing', 'No. 1, No. 2 and No. 3 Structural Light Framing', 'Construction, Standard and Utility Light Framing', and 'Stud Grade'. The stress grades available in the UK used to be a mix of No. 1 and No. 2, and a mix of Construction and Standard Light Framing. However, the UK structural design code BS5268: Part 2: 1988 now allocates the same stresses to No. 1 and No. 2. As a consequence, this mix is replaced simply by supply of No. 2, and Select Structural Light Framing is more readily available. No. 3, Utility and Stud Grade are rarely if ever supplied to

Table C5.2 Narrower sizes of CLS

Nominal size (inches)	Basic size when dry (mm)
2 × 2	38 × 38
2 × 3	38 × 63
2 × 4	38 × 89

See text for tolerances which are related to the basic size.

the UK. Three sizes are given in Table C5.2 but 38 × 89 mm and 38 × 63 mm are the sizes most readily available. Machine stress graded CLS is now being supplied to the UK to the NLGA Rules.

The tolerances on the basic sizes when dried to 19% moisture content are (according to BS4471: 1987 and the NLGA Rules): no minus tolerance, with plus tolerance unlimited. (The plus tolerance is usually small.)

Lengths are in 2 feet increments from 8 to 20 feet.

The cost of CLS sizes is often quoted as a price per 'Standard'. This refers to a Petrograd Standard which equals 4.671 m^3. In these cases the timber will normally be charged as the 'nominal' Imperial size in inches. For example 38 × 89 mm is normally charged as though it is 2 × 4 inch (50.8 × 101.6 mm).

SAWN SIZES OF STRESS GRADED TIMBER

Until a few years ago, in addition to the surfaced CLS described above, all stress graded sawn sizes exported to the UK from Canada were stress graded to the NLGA rules, and were cut to NLGA sizes and tolerances. Currently an increasing amount of sawn timber exported to the UK is sawn to the sizes and tolerances of BS4471: 1987. In either case, the timber is mainly Spruce-Pine-Fir with some supply of Hem-Fir, and it is normal for the timber to be supplied 'green' although there are discussions on dry sawn timber being supplied.

When timber to the NLGA sizes and tolerances is supplied, it is normal for the stress grading to be to the NLGA rules. When timber to the BS4471 sizes and tolerances is supplied, the stress grading is either to the NLGA rules or to the visual 'General Structural' or 'Special Structural' rules of BS4978: 1988 (see Chapter B3).

The NLGA stress grades available in the UK used to be mainly a mix of 'No. 1 and No. 2 Structural Joist and Plank'. However, the UK design code BS5268: Part 2: 1988 now allocates the same stresses to No. 1 and No. 2. As a consequence, as well as No. 2 being used separately, Select Structural Joist and Plank may be available more in NLGA sawn sizes, although the move towards more supply of BS4471 sizes and BS4978 stress grades should be taken into account.

The sizes and tolerances of BS4471: 1987 are given in Table B1.1. The NLGA sawn sizes most readily available in the UK are given in Table C5.3. The tolerances permitted on NLGA rough sawn green sizes are in accordance with Paragraph 747 of the NLGA (1987) rules. The minus tolerances vary from ⅛ inch to ¼ inch on sizes of 2 inch to 8 inch. Plus tolerances are twice the minus tolerances. The maximum minus tolerances are permitted on only 10% of pieces. A reduction of 2.5% in size is taken in estimating the minimum dry sizes given in Table C5.3. This is taken from BS4471: 1987 assuming a 10% moisture

Table C5.3 Sizes sawn to the NLGA rules and tolerances

Rough sawn 'green' size (inches)[1]	Minimum 'green' size (inches) deducting Para. 747 permitted minus tolerances[2]	Minimum size at 20% moisture content to the nearest mm assuming 2.5% drying reduction[3]
1⅞ × 4	1¾ × 3¹³⁄₁₆	43 × 94
× 6	× 5¹³⁄₁₆	× 144
× 8	× 7¾	× 192
× 10	× 9¾	× 241
× 12	× 11¾	× 291
1¾ × 4	1⅝ × 3¹³⁄₁₆	40 × 94
× 6	× 5¹³⁄₁₆	× 144
× 8	× 7¾	× 192
× 10	× 9¾	× 241
× 12	× 11¾	× 291

[1] Widths are normally in even inch widths, but sometimes 7 inch and 9 inch widths are available.
[2] See text.
[3] Note that the size to take in structural designs is normally slightly larger than these (see text).

content reduction from the 'green' size to the dry size at 20% moisture content, and a reduction of 1% in dimensions for each 4% reduction in moisture content.

Lengths are in 2 feet increments from 8 to 24 feet, or longer.

Normally with stress graded timber, it is necessary to base the design size on the minimum production size when 'dry', therefore sizes in the right hand column of Table C5.3 are calculated on that basis for a moisture content of 20%. One would normally expect the design size of sawn timber at 20% moisture content to be taken as being slightly larger than the minimum production size if only 10% of pieces are permitted to be as small as the minimum size (e.g. say 44 × 194 mm rather than 43 × 192 mm). See the UK design code for the actual rules.

MERCHANTABLE 'R LIST' TIMBER

A small amount of Canadian sawn timber is sold to the 'Export R List Grading and Dressing Rules' rather than the NLGA Rules. Relevant rough sawn 'green' sizes are given in Table C5.4. The timber is not stress graded in Canada but is sold as a mix of 'Selected Merchantable', 'No. 1 Merchantable', and 'No. 2 Merchantable'. Occasionally 'No. 3 Common' is available in the UK.

The species available are normally Spruce-Pine-Fir and Hemlock with some Douglas Fir-Larch and Western Red Cedar. They are normally sold undried ('green').

The tolerances permitted are in accordance with Paragraph 11 of the 'R List'. If R List timber is stress graded in the UK, the specifier should

Table C5.4 Sawn merchantable 'R List' sizes

Rough sawn 'green' size (inches)[1]
$1\frac{7}{8} \times 4$
$1\frac{7}{8} \times 6$
$1\frac{7}{8} \times 8$
$1\frac{7}{8} \times 10$
$1\frac{7}{8} \times 12$
3×4
3×6
3×8
3×10
3×12
4×4
4×6
4×8
4×10
4×12

[1] Widths are normally in even inch widths but sometimes 5 inch, 7 inch and 9 inch widths are available.

take note of the tolerances and make an allowance for drying in determining the size to use in designs.

Lengths are in 2 feet increments from 8–24 feet, or longer.

'DOOR STOCK'

'Door Stock' is sold in the species Douglas Fir-Larch or Hemlock (including Amabilis Fir). It is normally sold undried ('green'). To match the Canadian rules, 90% is required to be 'quarter-sawn' (i.e. with the growth rings meeting the faces at an angle of not less than 45°) or, to use the Canadian term, 'vertical grain'. The timber is cut away from the pith. The grain is straight.

Several sizes are cut for door stock but those exported to the UK are mainly of rough sawn green sizes 2 inch × 4 inch or 2 inch × 8 inch, plus some 2 inch x 6 inch material and a small volume of thicknesses of $1\frac{5}{8}$ inch.

The NLGA Rules (Paragraph 156) permit the widths *when dry* to be 'scant' by not more than $\frac{1}{8}$ inch in 4 inch and 6 inch widths, and $\frac{1}{4}$ inch scant in 8 inch widths. Thicknesses must be full size *when green*. Therefore, the minimum dry (20% moisture content) sizes in millimetres (allowing a 2.5% drying reduction from green) are (to the nearest millimetre):

50 × 98 mm	40 × 98 mm
50 × 149 mm	40 × 149 mm
50 × 197 mm	40 × 197 mm

Lengths are in 1 foot increments from 6 to 20 feet.

There are three grades called 'Door Cuttings' which are 'No. 1 Cuttings', 'No. 2 Cuttings' and 'No. 3 Cuttings', which are available in mixes known as 'Factory Select', 'No. 1 Shop (Door Stock)', 'No. 2 Shop (Door Stock)' and 'No. 3 Shop (Door Stock)'. These grades and combinations are described in detail in the NLGA Rules.

'CLEAR' GRADES OF DOUGLAS FIR-LARCH OR HEMLOCK

'Clear' grades of Douglas Fir-Larch or Hemlock (including Amabilis Fir) are taken from the outside of logs. There are three grades of 'Clears' which are 'No. 2 Clear and Better', 'No. 3 Clear' and 'No. 4 Clear'. The grades are described in the 'Export R List'. The lower qualities may have more numerous knots on the face than the higher qualities.

These Clears are normally sold undried ('green'). The sizes given in Table C5.5 are for green rough sawn timber 'undressed'. The permitted tolerances are detailed in Paragraph 11 of the 'Export R List'. The minus tolerances vary from $\frac{1}{16}$ inch to $\frac{1}{8}$ inch, plus tolerances from $\frac{1}{8}$ inch to $\frac{1}{4}$ inch on sizes up to 12 inch. The plus tolerances permitted are twice the minus tolerances. A reduction of 2.5% in size from green should be taken in estimating the dry sizes (at 20% moisture content).

Lengths are in 2 feet increments from 8–20 feet.

'CLEAR' GRADES OF WESTERN RED CEDAR

'Clear' grades of Western Red Cedar are taken from the outside of logs. The grades of 'Clears' are 'No. 2 Clear and Better', and 'No. 4

Table C5.5 Green sawn sizes of 'Clears' in Douglas Fir-Larch or Hemlock

'Clear' grades of Douglas Fir-Larch or Hemlock rough sawn 'green' size (inches)[1,2]		
2 × 4	3 × 4	4 × 4
2 × 6	3 × 6	4 × 6
2 × 8	3 × 8	4 × 8
2 × 10	3 × 10	4 × 10
2 × 12	3 × 12	4 × 12

[1] See text for tolerances.

[2] Thicknesses of $1\frac{3}{4}$ inch are sometimes available.

Table C5.6 Green sawn sizes of 'Clears' in Western Red Cedar

'Clear' grades of Western Red Cedar rough sawn 'green' size (inches)[1,2]			
1¾ × 4	2 × 4	3 × 4	4 × 4
1¾ × 6	2 × 6	3 × 6	4 × 6
1¾ × 8	2 × 8	3 × 8	4 × 8
1¾ × 10	2 × 10	3 × 10	4 × 10
1¾ × 12	2 × 12	3 × 12	4 × 12

[1] See text for tolerances.
[2] Occasionally widths of 5 inch, 7 inch, 9 inch and 11 inch are available.

Clear'. The grades are described in the 'Export R List'.

These Clears are normally sold undried ('green'). The sizes given in Table C5.6 are for green rough sawn timber 'undressed'. The permitted tolerances are detailed in Paragraph 11 of the 'Export R List'. The minus tolerances on thickness vary from $1/16$ inch for 2 inch thicknesses to ⅛ inch for 3 to 5 inch thicknesses. The minus tolerances on width vary from ⅛ inch for 4 inch widths to ¼ inch for 12 inch widths and wider. The plus tolerances permitted are twice the minus tolerances. A reduction of 2.5% in size from green should be taken in estimating the dry sizes (at 20% moisture content).

Lengths are in one foot increments from 8 to 20 feet.

'FACTORY LUMBER'

'Factory Lumber' is taken from a middle band between the centre of the tree where there is likely to be several knots, and the outer part which will be largely knot-free. Therefore, Factory Lumber is largely knot-free but is likely to contain a few large knots. The intention is that it should be re-processed in a manufacturer's factory to yield the clear timber between knots.

There are two types of Factory Lumber which are known as 'Shop Lumber' or 'Factory Flitches'. In the UK they are usually sold undried and in the sawn condition ('undressed').

'Shop Lumber' is usually available in Western Red Cedar. It is normally available in 'green' rough sawn thicknesses of 1¾ inch or 2 inch and in several rough sawn widths up to 12 inch. The intention is that it should be left in these thicknesses to be cross-cut and perhaps resawn (say two 4 inch widths from one 8 inch width) in obtaining clear timber (see Fig. C5.1).

'Factory Flitches' are usually in Hemlock, Douglas Fir-Larch, or Western Red Cedar in 'green' rough sawn thicknesses of 3 inch and 4 inch, and in 'green' rough sawn even widths from 4 inch to 12 inch. Pieces are initially cut mainly 'flat grain' (i.e. with the growth rings meeting the faces at an angle of less than 45°—see Fig. C5.1). The

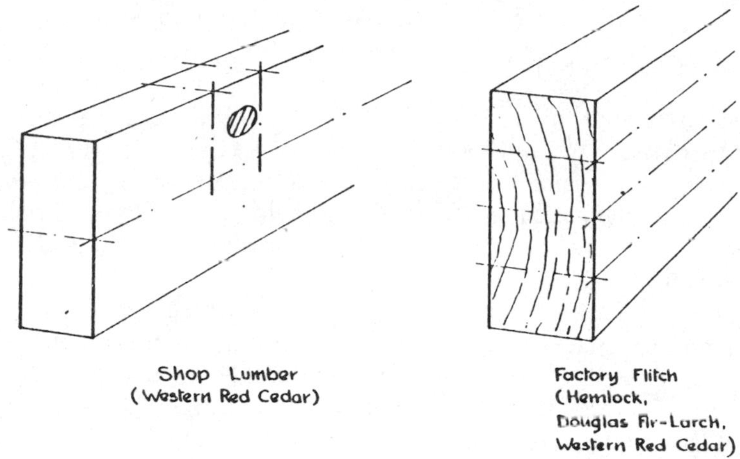

Fig. C5.1 Factory Lumber

intention is that a flitch should be resawn as shown in Fig. C5.1 to produce mainly vertical grain (edge grain) pieces.

The approximate tolerances on the 'green' sawn sizes are said to be as for 'Rough Clears' in Paragraph 11 of the 'Export R List' (see the notes immediately above for 'Clear' grades of Western Red Cedar).

Lengths are in 2 feet increments from 8 to 20 feet.

LARGE SIZED TIMBER

Large sizes of Canadian timber are available stress graded to the NLGA Rules or un-stress graded to the 'Export R List'. The sizes imported to the UK are normally un-stress graded to the 'Export R List'. The quality normally sold to the UK is a mixture of 'Selected Merchantable', 'No. 1 and No. 2 Merchantable' ('as developing' usually with not more than 25% of the mix being No. 2). 'Export R List' Large Sized Timbers have a thickness of at least 6 inch.

Large Sized Timbers stress graded to the NLGA rules are given the descriptions:

- *'Beams and Stringers'* (which are 5 inch thick or thicker, with the width more than 2 inch greater than the thickness) and
- *'Posts and Timbers'* (which are 5 inch × 5 inch or larger, with one dimension not more than 2 inch greater than the other).

Obviously, normally these terms have more generalized meanings in the UK and therefore care must be taken in using them. They are shipped undried ('green').

The approximate tolerances on the 'green' sizes are as Paragraph 11 of the 'Export R List' for timber supplied to the R List, and as Paragraph 747 of the NLGA rules for timber supplied to the NLGA rules.

The term 'Beams and Stringers' refers to sizes which are 'more than 2 inch off-square' (i.e. having one dimension more than 2 inch different to the other dimension). The most common size of 'Beams and Stringers' is 6 inch × 12 inch, with 16 inch being normally the largest width imported to the UK. Sizes are normally available in Douglas Fir-Larch and Hemlock.

The term 'Posts and Timbers' refers to sizes which are 'not more than 2 inch off-square'. The largest size is normally 16 inch × 16 inch (or 8 inch × 8 inch in Spruce-Pine-Fir). The smallest size is 5 inch × 5 inch. Common sizes in inches in Douglas Fir-Larch, Hemlock, or Western Red Cedar are 12 × 12, 10 × 12, 10 × 10 and 8 × 10.

Lengths are in 1 foot increments from 8 feet up to 32 feet, with some mills able to provide 40 feet lengths.

OTHER SPECIALITY ITEMS

Although other speciality items are available from Canadian mills, in the UK it is usual for a manufacturer to obtain them by re-sawing from larger sizes. For example, ladder stiles may be cut from 3 inch thick 'Clears'.

C6 UNITED KINGDOM FORESTS AND SAWMILLING PRODUCTION OF SITKA SPRUCE AND OTHER SOFTWOOD SPECIES

THE FORESTS

The mainland of the United Kingdom extends from latitude 50° North to latitude 58° North. The land area, including Northern Ireland, is 24 million hectares, of which 10% is estimated to be under forestry (not necessarily all productive).

The area of forest under the control of the Forestry Commission in England, Scotland and Wales is given as about 930 000 hectares in *Forestry Facts and Figures 1986-87*, published by the Forestry Commission. By far the largest area of the Forestry Commission forests is planted with conifer. The area of forest is being increased. The area of forest in Northern Ireland is 85 000 hectares.

The principal conifer is Sitka Spruce (*Picea sitchensis* (Bong.) Carr.) which is estimated to cover more than 400 000 hectares. In addition there is Norway Spruce (*Picea abies* (L.) Karst.), Scots Pine (*Pinus sylvestris* L.), Corsican Pine (*Pinus nigra Arnold var maritima* (Ait.) Melv.), Douglas Fir (*Pseudotsuga menziesii* (Mirb.) Franco), European Larch (*Larix decidua* Mill.) and Japanese Larch (*Larix kaempferi* (Lamb.) Carr.), plus a few others, and several hardwood species.

Note that, in the other chapters in this book, most of the references to a timber grown in the UK is to 'British-grown Sitka Spruce', which is the terminology used in structural design code BS5268: Part 2 for the most common of the species.

The estimate of average annual production of conifer 'roundwood' (i.e. softwood and hardwood logs) in England, Scotland and Wales over the period 1982–6 (by the UK Home-Grown Timber Advisory Committee) is 4 130 000 m^3 over bark which is expected to increase by some 50% by the 1992–6 period. Currently 70% of the production is from Forestry Commission forests. Forty-five per cent of the production from England, Scotland and Wales is estimated to be of the two spruces.

The estimate of annual production of conifer roundwood in Northern Ireland is approaching 80 000 m^3 over bark of which 70–75% is of the spruces.

The spruces reach maturity in the UK in 60–80 years, reaching a height of up to 30 m, and a reported diameter at breast height of as much as 600–900 mm over bark in the best plantations. However, economic rotations range from 45–60 years.

Approximately 47–50% of the roundwood production is 'small roundwood' (i.e. defined for Great Britain as 'logs of less than 140 mm diameter over bark plus 50% of the logs having a diameter of between 150 mm and 170 mm over bark').

PRODUCTION

Much of the production from UK forests is sold as roundwood. For example, large quantities of logs are sold for pulpwood or chippings, and the National Coal Board buys quantities to use as pit props or 'splits' (i.e. a log split into two segments). In *Forestry facts and figures 1986–87*, The Forestry Commission estimated that the total delivery of 'softwood sawlogs' in 1986 in Great Britain was 2 390 000 m^3 under bark. Assuming a conversion rate of 55% (allowing wane for lower quality uses) leads to an estimate of 1 310 000 m^3 of sawn softwood, including an estimate for the spruces of about 600 000 m^3.

The major uses for UK sawn softwood are for:

- box and pallet manufacture,
- fencing (Larch and Douglas Fir preferred),
- the construction industry,
- the furniture industry (which takes the top quality), and
- mining timber.

SAWMILLING

The frame saw is used in the UK but many of the mills which mainly saw spruce use profile chippers and band saws. A few mills use chamber (compartment) kilns, but often the sawn timber is sold undried ('green') or air dried depending on the intended use.

If timber (usually Sitka Spruce) is to be used stress graded, it is becoming more common for it to be machine stress graded to the M75 rules of BS4978: 1988, which lead to it having comparable strength and stiffness to imported European Whitewood stress graded to the

General Structural level of BS4978: 1988. As an alternative, it can be machine stress graded directly to one of the Strength Classes (normally SC3) of design code BS5268: Part 2: 1988 (see Chapter B3).

Other than for stress grading purposes, the bulk of UK softwood is sold either as 'run of the mill' (with poor looking pieces or those with rot or excessive wane excluded), or graded to a customer's specification (e.g. for pallets, fences or kitchen furniture).

The UK sawmilling industry tends to work to sizes tabulated in BS4471: 1987 but is also willing to cut to customer's specifications.

BRAZILIAN FORESTS AND SAWMILLING PRODUCTION OF PARANA PINE

C7

THE FORESTS

Brazil extends from latitude 5° North to latitude 34° South. The land area is 850 million hectares and about 40% is covered by forests. Of the estimated 335 million hectares of forested land less than 1% is covered with conifer. There are many species of hardwoods but this chapter deals only with the conifer softwood Parana Pine (*Araucaria angustifolia* (Bert.) O.Ktze.). Most of the commercial forests of Parana Pine occur in the southern areas of Brazil between latitudes 22° South and 28° South. The species grows in other South American countries but supply to the UK is from Brazil.

In the 1960s, concern was expressed at the lack of re-afforestation. Despite recent action on re-afforestation it seems clear that Parana Pine has been over-cut, and some mills which used to saw Parana Pine have now changed to sawing hardwoods. What is left of the old growth stands of Parana Pine is mostly in private ownership. The area of Parana Pine forest is reported to be about 250 000 hectares with almost twice that area containing mixed stands of Parana Pine and hardwoods. Also, alternative faster-grown species of pine are being grown in plantations in traditional Parana Pine regions.

The trees are straight and can reach a height of 35 m. Much of the trunk is free of branches in the standing tree. The diameter of mature trees is between 0.8 m and 1.5 m. Trees can reach a diameter of 500 mm in 50 years. The growth characteristics are such that a high percentage of trees contain compression wood (reaction wood) which has an influence on the distortion of the sawn timber on drying and re-sawing.

The production of sawn softwood has levelled-off at between 4 and 4.5 million m^3. The export used to be about one quarter of production but has been reduced to around 500 000 m^3 of which about 40% is exported to the Northern Hemisphere. The largest single export market is Argentina. Export of Parana Pine to the UK in 1979 was 85 000 m^3 which was just over 1% of the softwood imported. The export to the UK in 1987 was 42 000 m^3 which was about 0.5% of the softwood imported.

SAWMILLING AND SAWN TIMBER

Although there are over 2500 sawmills in Brazil, most deal mainly with hardwoods. Fewer than 500 handle Parana Pine. Parana Pine is sawn oversize in the 'green' condition so that it is the full basic sawn size when dried to around 20% moisture content. Band saws are used for the primary sawing, with circular saws used for edging (i.e. cutting to width).

A limited range of sizes are imported to the UK, mainly for stair strings and treads. The sizes (in mm) of main interest are:

25 × 200	25 × 225	25 × 250	25 × 275	25 × 300
	32 × 225	32 × 250	32 × 275	32 × 300
		38 × 250	38 × 275	38 × 300

Lengths of up to 5.4 m are available, although the most popular length with UK buyers is 4.2 m for stair strings, with 4.8 m being the usual maximum length available. Packages of single length such as 4.2 m are available at extra cost.

To satisfy the conditions of contract for shipment to the UK, the timber must be 'properly seasoned for shipment to the United Kingdom', and the official price list states that 'timber will be supplied chemically treated'. This is considered to be important. Parana Pine is shipped under deck and can deteriorate on the long sea journey if not sufficiently dried and protected. Until recent years, Parana Pine sold to the UK was air dried, although all large mills have kilns. Parana Pine is available kiln dried to around 18% from Brazilian mills and it is also possible to obtain it dried and planed. There are Brazilian government incentives for companies which sell timber planed and dried and, as a consequence, kiln dried and planed goods now dominate the import of Parana Pine to the UK.

The mill grading rules for Parana Pine are set by the Brazilian government. There are four grades of which the lowest may be considered to be similar to Nordic 'sixth' quality, but only Grade 1 and Grade 2 are exported to the UK. Small sections are sold to the UK as '1st Quality' (i.e. Grade 1), but the usual commercial grade exported to the UK is a mixed grade called 'Prime Grade'. Prime Grade consists of 80% Grade 1 and 20% Grade 2.

Grade 1 permits some wane on one edge of sawn timber, not planed timber, and small knots on one face.

Grade 2 permits some wane on both edges of sawn timber, not planed timber, knots on both faces, and a few worm holes.

The shipping marks are given in *Shipping marks on timber*, by Benn Publications Ltd. (currently out of print).

C8 PORTUGUESE FORESTS AND SAWMILLING PRODUCTION OF MARITIME PINE

THE FORESTS

Portugal extends from latitude 37° North to latitude 42° North. The land area is 9.2 million hectares. There are 3.1 million hectares of forest of which some 1.3 million are of Maritime Pine.

The total volume of conifer standing timber is estimated to be 92 million forest cubic metres over bark, and the annual increment of growth is estimated at 7.3 million forest cubic metres over bark. It is reported that the Portuguese Government's 'Forestry Action Programme' will lead to an increase of 400 000 hectares of Maritime Pine forests.

Maritime Pine (*Pinus pinaster* Ait.) in Portugal can have a diameter at breast height of up to 1 m, but most trees felled for timber have a diameter at breast height of between 500 and 750 mm. Trees mature at 70–80 years (although felled for some sawn timber uses at an earlier age), and can remain healthy if left until an age of about 130 years, after which some deterioration begins.

SAWMILLING AND SAWN TIMBER

The import of Maritime Pine to the UK in 1987 was 631 000 m^3 which represented nearly 8% of the import. The figures for 1986 were 635 000 m^3 and 9%.

There are some 1200 sawmlls. Some mills arrange export themselves, whilst many of the others supply 'quay exporters' who export for several mills. The shipping marks appear in *Shipping marks on timber*, by Benn Publications Ltd. (currently out of print).

The sawing equipment is mainly band saws, with circular saws used for trimming.

There are a few kilns and some of the timber is air dried but the timber exported to the UK is usually 'green' and treated with anti-stain.

There are no published grading rules but the sawn timber is sold as being suitable for pallet boards, fencing, box-boards, with some sizes used for packaging material. Generally the grading qualities are accepted as being suitable for the purpose for which they are intended. As far as this book is concerned, the main use of Maritime Pine is for fences.

For pallet boards, within the limits of 75–150 mm in width and 12–100 mm in thickness, mills are able to cut sizes as required. The normal lengths are between 0.75 m and 2.5 m with 3 m being the usual maximum. It is possible, but not easy, to obtain slightly longer lengths.

For boxboards, thicknesses vary from 4–12 mm, widths from 50–120 mm, in lengths from 0.1–0.5 m.

For fencing boards, thicknesses vary from 5–19 mm, widths from 30–125 mm, in lengths from 1.2–1.8 m.

The size of packages varies with the size of timber being shipped. The volume in each can be quite small (e.g. one cubic metre).

Virtually all packages are square both ends and are not normally wrapped.

GLOSSARY OF TERMS AND ABBREVIATIONS

A glossary and explanation of terms and abbreviations related to timber which are used in the text and which may not be familiar to readers. The explanations given are not necessarily official or BSI definitions but are intended to explain the normal uses of the terms:

anti-stain treatment A chemical solution applied to timber by immersion or enclosed spray at some sawmills, to minimize staining of the timber during initial transit and storage.

basic length As intended in BS4471: 1987, basic length is the length to which tolerances apply.

basic sawn size As intended in BS4471: 1987, basic sawn size is the size on which tolerances apply at 20% moisture content.

basic strength The strength of defect-free timber. (A term not used to the same extent as previously. The equivalent term used in BS5268: Part 2: 1988 is 'clear wood stress'.)

blue stain A discoloration caused by timber having been at a moisture content of about 20/22% or more for a period probably of at least a few weeks. It is not considered by BS4978: 1988 to be a structural defect but can detract from appearance. Also see 'sap stain'.

bow Lengthwise distortion on the weaker axis of a piece of rectangular timber. If there is distortion on either axis of square timber, it is normal to refer to it as bow rather than as spring (see 'spring').

BR (in relation to adhesives) Boil-Resistant (see Chapter B9).

bracking A timber trade term for grading, usually sawmill grading.

bright timber A term normally intended to refer to timber which has not been discoloured (e.g. without blue stain).

BWF The British Woodworking Federation.

cant A Canadian term for the centre part of a log after outside parts have been removed during conversion. (Nordic term 'centre-cut'.)

case-hardening A condition in timber in which the outer layers near the surface have dried before the centre, and have become set to such an extent as to cause stress between the outer and inner layers (see Chapter B2 including Fig. B2.2). Despite the wording, there is no hardening of the surfaces.

CCA Copper/chromium/arsenic timber preservative (see Chapter B5).

centre-cut A Nordic term for the centre part of a log after the first sawing operation to remove the outer parts (see Fig. C1.1).

chamber kiln (compartment kiln) A drying kiln where the timber remains in one position during the drying cycle (as opposed to a progressive kiln).

check Separation of fibres along the grain forming a crack or fissure that does not extend through timber from one surface to another.

Clears A term for a sawmill grade, mainly applied to Canadian timber, indicating that the timber is either free of knots, or has useful lengths of timber between knots.

CLS Canadian Lumber Standards sizes (see Chapters C5 and B3).

compression wood Abnormal wood (reaction wood) formed in the growing tree typically underneath branch positions and in leaning or crooked trunks of softwood (coniferous) trees. Zones of compression wood are typically denser and darker than the surrounding normal wood and have abnormally high longitudinal shrinkage on drying, which can lead to distortion and splitting.

Concealed surface As intended by BS1186, a surface in joinery or wood trim which, after installation in the building, will be concealed not merely by decoration.

conditioning A stage at the end of the drying cycle, or after kiln drying of timber, at the sawmill or during a secondary drying process, during which any moisture gradient in the timber is reduced or eliminated as the timber adjusts to the equilibrium moisture content governed by the ambient temperature and relative humidity.

cross-cut A cut across the grain (e.g. to cut one length into two or more lengths).

cup Curvature of a piece of timber across the width of a face.

deep sawn or deeping Re-sawing timber lengthwise parallel to the faces (i.e. parallel to the wider dimension).

dimensioned A North American term referring to timber which has been processed to make the cross-sectional dimensions uniform. Also see 'dressed' or 'surfaced' timber.

door stock Timber cut and graded at a sawmill for the production of a panel door (see Fig. A2.1 and Chapter C5).

dressed (or 'surfaced') A North American term referring to timber which has been processed by a planing machine 'for purposes of attaining smoothness of surface and uniformity of size', on one side (i.e. face), two sides, one edge, two edges, or a combination of sides and edges.

dry As intended by structural design code BS5268: Part 2 for allocating permissible stresses, 'dry' refers to a moisture content not in excess of 18%.

dry As intended by UK codes of practice on preservation (see

Chapter B5), 'dry' refers to service moisture contents generally 'below about 20%' (see Clause 3.2a of BS5268: Part 5: 1989).

Durable As intended by BRE and British Standards, a specific category of durability of timber in contact with the ground (see Table B5.1).

edge Either of the two narrower surfaces of a piece of rectangular timber.

edge grain timber (or quarter-sawn timber) Timber converted so that the growth rings meet the face at an angle of not less than 45° (see Figs B2.1 and C5.1).

equilibrium moisture content The moisture content at which timber neither gains nor loses moisture when exposed to a given constant condition of temperature and relative humidity.

ER (in relation to preservation) Extremely Resistant (see below).

ex A timber term meaning 'from' or 'out of' (e.g. 35 × 95 mm 'ex' 38 × 100 mm).

Exposed surface As intended by BS1186, a surface in joinery or wood trim which is 'Exposed' in the final work. Such a surface is considered to be 'Exposed', even if finished with an opaque paint.

Extremely Resistant A term with a specific meaning in relation to the amenability to preservation of the sapwood or heartwood of a timber species (see Table B5.3 and associated notes in Chapter B5).

face Either of the two wider surfaces of a rectangular timber section, or any of the four surfaces of square timber.

Factory Flitch A term specific to large sawn sizes of North American timber intended to be re-sawn in a specific way (see Fig. C5.1).

Factory Lumber Shop Lumber or Factory Flitches (see Fig. C5.1).

fibre saturation point The moisture content of timber at which all 'free' moisture has been removed from the cell cavities but at which the cell walls are still saturated with 'bound' water. It is only when the moisture content of timber falls below this level that timber begins to shrink (see Chapter B2).

fifth quality A sawmill grade normally associated with Nordic, Russian and European timber. When comparing Russian and Nordic timber, it should be realized that the term has a different significance to each. For Russian timber, broadly speaking fifth quality is equivalent to Nordic sixth quality (see Chapters C1, C2, C3 and C6).

figure The appearance of the grain pattern on the surface of timber.

fissures A term including checks, splits and shakes.

flat grain Timber converted so that the growth rings meet the face at an angle of less than 45° (see Fig. C5.1).

fourth quality A sawmill grade normally associated with Russian and Czechoslovakian timber (see Chapters C2 and C4). Broadly speaking, Russian fourth quality is equivalent to Nordic unsorted.

frame saw A type of sawing frame normally containing several saws, normally used more at a sawmill than by an importer or manufacturer. A frame saw gives a distinctive appearance to sawn faces. (Canadian term 'gang saw'.)

gang saw See 'frame saw'.

FTS 'Free Trafalgar Square'. A term with some exclusivity to Norwegian Spruce trees, used in December only once a year.

green timber Freshly-felled timber or timber still containing free moisture in its cell cavities (i.e. timber with a moisture content at or above the fibre saturation point).

growth stresses Stresses built into the wood during the growth of the tree. Such stresses are often released during conversation of the logs to sawn timber, or during drying, or during subsequent re-sawing, and may lead to distortions in the timber.

hardwood By convention, timber of broad-leaved trees (belonging to the botanical group Angiosperms). It should be realized that the term hardwood is a botanical grouping not based on mechanical properties such as hardness or softness.

heartwood Wood in the inner annular rings (growth rings) which, in the growing tree, has ceased to contain living cells and in which the reserve materials (e.g. starch) have been removed or changed into more durable substances.

high-porosity See 'over-porosity'.

in the white Unpainted joinery.

juvenile wood Wood immediately around the pith of a tree, log or sawn timber formed in the early years of growth of the tree.

KAR Knot area ratio (see below).

Knot area ratio A method of defining the way in which knots affect the cross-sectional area of a piece of timber (see Fig. B3.2).

knot-free To a sawmiller, knot-free means that two edges and *one* face of a piece of timber are free of knots (therefore this is a term to use only with care).

knot types As defined in BS6100: Section 4.1:

A *bark-ringed* knot or an *encased* knot is a knot wholly or partially surrounded with bark or resin.

A *dead knot* is a knot having fibres intergrown with those of the surrounding wood to the extent of 25% or less of the cross-sectional perimeter.

An *intergrown knot* or *live knot* is a knot having fibres intergrown with those of the surrounding wood to the extent of approximately 75% or more of the cross-sectional perimeter.

A *knot cluster* is a group of knots around which the fibres are deflected.

A *loose knot* is a dead knot that is not held firmly in place.

A *sound knot* is a knot free from rot, solid across its face and at least as hard as the surrounding wood.

A *tight knot* is a knot held firmly in place. (Note that there is no requirement for the extent to which the fibres must be intergrown with the surrounding wood for the knot to qualify as 'tight', therefore a *dead knot* can be described as a tight knot.)

An *unsound knot* is a knot not solid across its face, or one which is softer than the surrounding wood due to rot or other defects.

laminate One length of solid timber in a structural glulam member, or a member laminated for joinery.

made good A term usually referring to repairing an item of joinery or wood trim by a plug, insert or filler.

margin In relation to grading of timber, a margin is a portion of the cross-section of a piece of solid timber near either edge. In the specific case of stress grading to BS4978, a margin is the outer quarter of the cross-sectional area (see Fig. B3.4).

margin condition In the specific case of stress grading visually to BS4978, a margin condition is said to exist when more than one half of the area of one margin is occupied by the projected area of knots.

Moderately Durable As intended by BRE and British Standards, a specific category of durability of timber in contact with the ground (see Table B5.1).

Moderately Resistant A term with a specific meaning in relation to the amenability to preservation of the sapwood or heartwood of a timber species (see Table B5.3 and associated notes in Chapter B5).

moisture content (expressed as a percentage)

$$\frac{\text{weight of moisture in the timber} \times 100\%}{\text{weight of oven-dried timber}}$$

moisture gradient A variation in moisture content between the outer and inner layers at a cross-section. If a gradient exists in newly dried timber, normally the outer surfaces are drier than the centre of the piece.

MR (in relation to adhesives) Moisture Resistant (see Chapter B9).

MR (in relation to preservation) Moderately Resistant (see above).

NASC National Association of Scaffolding Contractors.

NHBC The National House-Building Council.

NLGA The Canadian National Lumber Grades Authority.

nominal size The term nominal size is deprecated by BSI but is still used within the timber trade. Specifiers and non-specialists in timber are advised to find out exactly what is intended each time that the phrase is used. Sometimes the term can refer to a dried size, sometimes to a size before drying. As far as sawn timber is concerned, nominal size normally refers to the size to which tolerances apply, but the tolerances may be larger than those of BS4471. As far as processed timber is concerned, sometimes nominal size can refer to the sawn size from which the timber has been processed but, for example with CLS (see Chapter C5), nominal size can have a significantly different meaning.

Non-Durable As intended by BRE and British Standards, a specific category of durability of timber in contact with the ground (see Table B5.1).

normal bracking The quality of grading normally associated with a particular sawmill.

notional charring rate The rate at which a timber member of a particular species over a certain defined size chars when fully exposed to a standard fire test (see Chapter B6).

over-porosity A condition which can occur in patches on the surfaces of sawn timber of certain species (e.g. European Redwood) if the logs from which the timber was cut were stored in water for several weeks in warm weather. Extensive sprinkling of logs with water for a long period in warm weather can also lead to over-porosity in the sawn timber. Over-porosity can lead to an unacceptable appearance if the timber is finished with a decorative stain (see Chapter B7).

P (in relation to preservation) Permeable (see below).

package A package of timber. The majority of packages are not more than 1.2 m wide and not more than 1.0 m high (i.e. a quarter-container module), banded, containing timber of the same size, and the same or different lengths (see Chapters C1–C8).

parcel of timber As intended by BS4471, a quantity of sawn timber (and presumably processed timber) of the same basic size, quality and description. The minimum number of pieces in a parcel is not defined.

Perishable As intended by BRE and British Standards, a specific category of durability of timber in contact with the ground (see Table B5.1).

Permeable A term with a specific meaning in relation to the amenability to preservation of the sapwood and heartwood of a timber species (see Table B5.3 and associated notes in Chapter B5).

pith The central core of a stem, tree or log, consisting mainly of soft tissue. Obviously it can occur in or on the surface of some pieces of timber.

ponding Storing logs in water. Also see 'over-porosity'.

processing Planing, moulding, regularizing, surfacing, sanding etc. of timber.

progressive kiln A drying kiln in which stacks of timber are moved in a straight line from the entry position to the exit position.

quarter-sawn timber See 'edge grain timber'.

R (in relation to preservation) Resistant (see below).

regularizing As intended by BS4471 for solid timber, a process for making the width (depth) of sawn timber more uniform by sawing or machining on one or both opposed edges.

regularized As intended by BS4169 for glulam, a glulam member having at least 50% of the surface area sawn or planed to remove the protruding laminations.

relative humidity The amount of moisture in the air compared to that at complete saturation of the air (i.e. the dew point) at a specific temperature.

resawn See Fig. B1.1.

resawn ex larger As intended by BS4471, a term to indicate that the size being quoted has been produced by being resawn from a larger size and consequently is permitted to have re-sawing minus tolerances. (It is worth noting that, at the time of specifying or quoting, often the potential supplier is not aware if a basic size or a basic size 'resawn ex larger' will be supplied.)

Resistant A term with a specific meaning in relation to the amenability to preservation of the sapwood or heartwood of a timber species (see Table B5.3 and associated notes in Chapter B5).

roundwood Felled trees or logs.

sapwood Wood in the outer annular rings (growth rings) inside the bark and cambium which, in the growing tree, contains living cells and reserve materials (e.g. starch) and conducts sap whilst the tree is growing.

sap stain A discoloration of timber resulting from the growth of certain fungi that derive nourishment from the cell contents but do not cause decomposition of the timber. It is principally confined to the sapwood. (Also see 'blue stain'.)

saw-falling quality A term to describe a sawmill mixed grade. It can indicate that a percentage of the higher qualities are included, but normally indicates that most of the volume is in lower qualities. When encountering the term for the first time in relation to the grading from a specific mill, it is prudent to check what quality mix is intended.

sawn size Timber sawn to size (normally rectangular or square) with or without wane, but not planed or otherwise processed.

scant A term sometimes used to describe timber smaller than the basic size less the permitted minus tolerances, or to describe the permitted minus tolerances.

schaal board A term used to describe an edge board of Nordic timber with a considerable amount of wane. The intention is that the board should be moulded by the buyer to eliminate the wane.

seasoned timber Dried timber, usually intended to mean air dried. There is no intention that the term should indicate that the timber has been dried for one or more seasons of the year.

Semi-Concealed Surface As intended by BS1186, a surface in joinery which, when an item such as a window or drawer is closed, is 'Concealed' not merely by decoration, but is 'Exposed' when the item is open.

service moisture content The moisture content, normally given as a range of moisture contents, at which timber will equalize after being left for a sufficient time in a specific service condition of temperature and relative humidity.

SFTC The Swedish Finnish Timber Council.

shake A separation of fibres along the grain, irrespective of the extent of penetration.

shipping dry timber Timber sufficiently dried to prevent deterioration in transit. The term normally applies mainly to the first transit overseas from a sawmill to the first purchaser such as an importer.

shipping mark A mark normally applied with a paint or dye on one or both ends of every piece of timber, to identify (by cross-reference to published lists) the grading sawmill or company and the quality of the piece.

shipping season A term used to state the months of the year during which timber can be shipped (i.e. delivered by sea) from a sawmill or sawmills in a particular area. When timber was exported to the UK mainly by ships which could not operate in the ice which often surrounds countries with very cold climates, the shipping season had great significance for the stocks which had to be held by importers. Now that powerful ships are used which can operate, for example in Baltic ice after passages have been cleared by ice-breakers, and/or timber is taken by rail or road to ice-free ports, the term is used in the UK far less than in the past and has less significance (but see 'Sawmilling and Sawn Timber' in Chapter C2).

Shop Lumber A term specific to large sawn sizes of North American timber intended to be re-sawn in a specific way (see Fig. C5.1).

side board Board sawn from the outer portion of a log.

sixth quality A sawmill grade normally associated with Nordic and some Polish timber (but not Russian). See Chapters C1 and C3.

softwood By convention, timber of coniferous trees (belonging to the botanical group Gymnosperms). It should be realized that the term softwood is a botanical grouping not based on mechanical properties such as softness or hardness.

split A separation of fibres along the grain forming a crack or fissure that extends through timber from one surface to another.

spring Lengthwise distortion on the strong axis of a piece of rectangular timber. If there is distortion on either axis of square timber, it is normal to refer to it as bow rather than spring.

Standard A 'Petrograd Standard Hundred' equivalent to 4.671 cubic metres of timber.

surfaced See 'dressed'.

thickness The thinner dimension of a piece of rectangular solid timber. In the case of a square section, it is normal to refer to one dimension as the thickness and to the other as the width.

TRADA The Timber Research and Development Association.

truck-bundle A parcel containing timber of differing lengths or of a limited number of different lengths (i.e. not square-both ends).

twist Lengthwise distortion on both axes of a piece of timber.

unsorted A sawmill grade normally associated with Nordic, Russian and other European timber. When comparing Russian and Nordic timber, it should be realized that the term unsorted has a similar but not identical meaning in each case (see Chapters C1 and C2).

usual bracking See 'normal bracking'.

vertical grain Another term (more used in North America than in Europe) for 'edge grain timber' or 'quarter-sawn timber' (see 'edge grain timber').

Very Durable As intended by BRE and British Standards, a specific category of durability of timber in contact with the ground (see Table B5.1).

wane The original rounded surface of a tree remaining with or without bark, on any face or edge of solid timber.

WBP (in relation to adhesives) Weather-proof and boil-proof (see Chapters B9 and B10).

width The larger dimension of a piece of solid timber. In the case of a square section, it is normal to refer to one dimension as the width and to the other as the thickness.

wet condition As intended by structural design code BS5268: Part 2 for allocating permissible stresses, 'wet' refers to a moisture content in excess of 18%.

wet condition As intended by the codes of practice on preservation related to the 'risk' of decay, 'wet' refers to service moisture contents generally in excess of about 20% (see Clause 3.2a of BS5268: Part 5: 1989).

yard blue Blue stain or sap stain caused by the moisture content of timber being at 20/22% or more for a few weeks subsequent to the timber having left the sawmill 'bright'.

REFERENCE DOCUMENTS

The following publications are referred to in this book.

BRITISH STANDARDS AND CODES

BS144: 1973 *Specification for coal tar creosote for the preservation of timber.*

BS459: 1988 *Specification for matchboarded wooden door leaves for external use.*

BS476: ---- *Fire tests on building materials and structures.*

BS476: Part 4: 1970 (1984) *Non-combustibility test for materials.*

BS476: Part 5: 1979 *Method of test for ignitability.*

BS476: Part 6: 1981 *Method of test for fire propagation for products.*

BS476: Part 7: 1987 *Method for classification of the surface spread of flame of products.*

BS476: Part 8: 1972 *Test methods and criteria for the fire resistance of elements of building construction.*

BS476: Part 11: 1982 *Method for assessing the heat emission from building materials.*

BS476: Part 20: 1989 *Method for determination of the fire resistance of elements of construction (general principles).*

BS476: Part 21: 1987 *Methods for determination of the fire resistance of loadbearing elements of construction.*

BS476: Part 22: 1987 *Methods for determination of the fire resistance of non-loadbearing elements of construction.*

BS476: Part 23: 1987 *Methods for determination of the contribution of components to the fire resistance of a structure.*

BS584: 1967 (1980) *Specification for wood trim (softwood).* It is the intention that this standard will be withdrawn to coincide with the publication of BS1186: Part 3.

BS585: Part 1: 1989 *Wood stairs. Specification for stairs with closed risers for domestic use, including straight and winder flights and quarter or half landings.*

BS585: Part 2: 1985 *Wood stairs. Specification for performance requirements for domestic stairs constructed of wood-based materials.*

BS644: Part 1: 1989 *Wood windows. Specification for factory assembled windows of various types.*

BS729: 1971 (1986) *Specification for hot dip galvanized coatings on iron and steel articles.*

BS745: 1969 *Specification for animal glue for wood.*

BS913: 1973 *Specification for wood preservation by means of pressure creosoting.*

BS1186: Part 1: 1986 *Timber for and workmanship in joinery. Specification for timber.*

BS1186: Part 2: 1987 *Timber for and workmanship in joinery. Specification for quality of workmanship.*

BS1186: Part 3: 19–– *Timber for and workmanship in joinery. Specification for wood trim: timber species, classification, workmanship in fixing, and standard profiles.* Under preparation, publication expected in 1990.

BS1202: Part 1: 1974 *Specification for nails. Steel nails.*

BS1204: Part 1: 1979 *Synthetic resin adhesives for wood. Specification for gap-filling adhesives.*

BS1204: Part 2: 1979 *Synthetic resin adhesives for wood. Specification for close-contact adhesives.*

BS1210: 1963 *Specification for wood screws.*

BS1282: 1975 *Guide to the choice, use and application of wood preservatives.*

BS1297: 1987 *Specification for tongued and grooved softwood flooring.*

BS1444: 1970 *Cold-setting casein adhesive powders for wood.* Withdrawn by BSI.

BS1579: 1960 *Specification for connectors for timber.*

BS1722: Part 5: 1986 *Fences. Specification for close boarded fences.*

BS1722: Part 6: 1986 *Fences. Specification for wooden palisade fences.*

BS1722: Part 7: 1986 *Fences. Specification for wooden post and rail fences.*

BS1722: Part 11: 1986 *Fences. Specification for woven wood and lap boarded panel fences.*

BS2482: 1981 *Specification for timber scaffold boards.*

BS3051: 1972 *Specification for coal tar creosotes for wood preservation (other than creosotes to BS144).*

BS4071: 1966 *Specification for polyvinyl acetate (PVA) emulsion adhesives for wood.*

BS4072: Part 1: 1987 *Wood preservation by means of copper/chromium/arsenic compositions. Specification for preservatives.*

BS4072: Part 2: 1987 *Wood preservation by means of copper/chromium/arsenic compositions.*

BS4169: 1988 *Specification for manufacture of glued-laminated timber structural members.*

BS4190: 1967 *Specification for ISO metric black hexagon bolts, screws and nuts.*

BS4320: 1968 *Specification for metal washers for general engineering purposes. Metric series.*

BS4471: 1987 *Specification for sizes of sawn and processed softwood.*

BS4787: Part 1: 1980 (1985) *Internal and external wood doorsets, door leaves and frames. Specification for dimensional requirements.*

BS4978: 1988 *Specification for softwood grades for structural use.*

BS5082: 1986 *Specification for water-borne priming paints for woodwork.*

BS5268: Part 2: 1988 *Structural use of timber. Code of practice for permissible stress design, materials and workmanship.*

BS5268: Part 3: 1985 *Structural use of timber. Code of practice for trussed rafter roofs.*

BS5268: Part 4: Section 4.1: 1978 *Structural use of timber. Fire resistance of timber structures. Recommendations for calculating fire resistance of timber structures.*

BS5268: Part 4: Section 4.2: 1989 *Structural use of timber. Fire resistance of timber structures. Recommendations for calculating fire resistance of timber stud walls and joisted floor constructions.*

BS5268: Part 5: 1989 *Structural use of timber. Code of practice for the preservative treatment of structural timber.*

BS5268: Part 6: Section 6.1: 1988 *Structural use of timber. Code of practice for timber frame walls. Dwellings not exceeding three storeys.*

BS5268: Part 7 *Structural use of timber. Recommendations for the calculation basis for span tables.*

Section 7.1: 1989 Domestic floor joists.
Section 7.2: 1989 Joists for flat roofs.
Section 7.3: 1989 Ceiling joists.
Section 7.4: 1989 Ceiling binders.
Section 7.5: 19— Rafters.
Section 7.6: 19— Purlins supporting rafters.
Section 7.7: 19— Purlins supporting sheeting.

(Sections 7.5–7.7 are under preparation, publication expected in 1990.)

BS5291: 1984 *Specification for manufacture of finger joints of structural softwood.*

BS5358: 1986 *Specification for solvent-borne priming paints for woodwork.*

BS5395: Part 1: 1977 (1984) *Stairs, ladders and walkways. Code of practice for the design of straight stairs.*

BS5442: Part 3: 1979 *Classification of adhesives for construction. Adhesives for use with wood.*

BS5534: Part 1: 1978 (1985) *Slating and tiling. Design* (including AMD3554 and AMD5781).

BS5588: Section 1.1: 1984 *Fire precautions in the design and construction of buildings. Residential buildings. Code of practice for single-family dwelling houses.*

BS5589: 1989 *Code of practice for preservation of timber.*

BS5707: Part 1: 1979 *Solutions of wood preservatives in organic solvents. Specification for solutions for general purpose applications, including timber that is to be painted.*

BS5707: Part 2: 1979 (1986) *Solutions of wood preservatives in organic solvents. Specification for pentachlorophenol wood preservative solution for use on timber that is not required to be painted.*

BS5707: Part 3: 1980 *Solutions of wood preservatives in organic solvents. Methods of treatment.*

BS5973: 1981 *Code of practice for access and working scaffolds and special scaffold structures in steel.*

BS6100: Section 4.1: 1984 *Glossary of building and civil engineering terms. Forest products. Characteristics and properties of timber and wood based panel products.*

BS6100: Section 4.4: 1985 *Glossary of building and civil engineering terms. Forest products. Carpentry and joinery.*

BS6375: Part 1: 1983 *Performance of windows. Classification for weathertightness (including guidance on selection and specification).*

BS6446: 1984 *Specification for manufacture of glued structural components of timber and wood based panel products.*

BS8103: Part 3: (anticipated for publication in 1990) *The structural design of low-rise buildings. Code of practice for timber floors and roofs in housing.*

BS8201: 1987 *Code of practice for flooring of timber, timber products and wood based panel products.*

BS8212: 1988 *Code of practice for dry lining and partitioning using gypsum plasterboard.*

BS Draft for Development DD74: 1981 *Performance requirements and test methods for non-structural wood adhesives.* (To be replaced by EN 204.)

STANDARDS OF COUNTRIES OTHER THAN THE UK

German DIN Standard 68 602: 1979 *Evaluation of adhesives for jointing of wood and derived timber products. Stress groups. Strength of bond.*

SIS 23 27 11: 1970 *The Swedish Standard for sawn and resawn timber.*

SFS 2511: 1982 The Finnish Standard *Sawn softwood dimensions.*

NS 3042: 1970 The Norwegian Standard *Sawn timber. Dimensions.*

National Lumber Grades Authority (NLGA) Standard Grading Rules for Canadian Timber. 1987.

Export R List Grading and Dressing Rules. 1951 (Revised 1971).

ECE recommended standard for stress grading and finger jointing of structural coniferous sawn timber. 1989.

EN 204, *Evaluation of non-structural adhesives for joinery of wood and timber drived products.* (In the course of preparation.)

PUBLICATIONS OF THE BUILDING RESEARCH ESTABLISHMENT

A handbook of Softwoods. 1977.

BRE Information Sheet IS22/78, *Machine re-grading: Why pieces can change grade.*

BRE Technical Note 24, *Preservative treatments for external softwood joinery timber*. 1982 (withdrawn).

BRE Digest 340, *Choosing wood adhesives.* 1989.

BRE Digest 175, *Choice of glues for wood.* 1975 (withdrawn with the publication of Digest 340).

BRE Digest 209, *Polyvinyl acetate adhesives for wood.* 1978 (withdrawn with the publication of Digest 340).

Timber drying manual. 1986.

BUILDING REGULATIONS AND APPROVED DOCUMENTS

The Building Regulations 1985 for England and Wales, including Amendments 1988 to the Manual and Approved Documents.

The Building Standards (Scotland) Regulations 1981.

Building Regulations (Northern Ireland) 1977.

Approved Document A1/2. 1985.

Approved Document B. 1985.

Approved Document to support Regulation 7 of The Building Regulations 1985 for England and Wales. 1985.

The Control of Pollution Act 1974.

The Control of Pesticides Regulations 1986.

PUBLICATIONS OF THE NATIONAL HOUSE-BUILDING COUNCIL

NHBC Practice Note 5 (1982), *Timber framed dwellings and external timber framed wall panels in masonry cross wall dwellings.*

Registered house-builder's handbook. 1985.

COMMODITY SPECIFICATIONS OF THE BRITISH WOOD PRESERVING ASSOCIATION

C1 (1986) Preservative treatment of timber to be used as packing in cooling towers.

C2 (1986) Preservative treatment of timber for use permanently or intermittently in contact with sea or fresh water.

C3 (1986) Preservative treatment of fencing timber.

C4 (1986) Preservative treatment of agricultural and horticultural timbers.

C5 (1986) Preservative treatment of non load-bearing external softwood joinery and external fittings (excluding cladding) not in ground contact.

C6 (1986) Preservative treatment of external timber cladding.

C7 (1986) Preservative treatment for timber for use in buildings in termite infested areas.

C8 (1986) Preservative treatments for constructional timbers (excluding walls of timber frame houses).

C9 (1986) Preservative treatments for timber frame housing.

OTHER PUBLICATIONS

Shipping marks on timber. Benn Publications Ltd. (Out of print since 1979.)

Handbook of the northern wood industries. The Swedish Timber and Wood Pulp Journal. 1986/87.

Guiding principles for grading of Swedish sawn timber for export. The Swedish Timber and Wood Pulp Journal. 1980.

The Finnish 'Grading rules for export timber'. The Association of Finnish Sawmillmen. 1979.

The Russian 'Northern Bracking' rules. 1985.

Scaffold board specification 1987. UK National Association of Scaffolding Contractors.

Forestry facts and figures 1986–87. The UK Forestry Commission.

DOE Waste Management Paper No. 16 *Wood preserving wastes.* 1980.

Performance in fire. The Swedish Finnish Timber Council. 1989.

Timber designers' manual, Baird and Ozelton. BSP Professional Books. 1986.

Draft of Eurocode 5: *Common unified rules for timber structures.* 1987.

Fire resisting door assemblies. The British Woodworking Federation. 1989.

British Columbia Forest Industry Fact Book 1988. Council of the Forest Industries of British Columbia.

INDEX